Spin Structure of the Nucleon

NATO Science Series

A Series presenting the results of scientific meetings supported under the NATO Science Programme.

The Series is published by IOS Press, Amsterdam, and Kluwer Academic Publishers in conjunction with the NATO Scientific Affairs Division

Sub-Series

I. **Life and Behavioural Sciences**	IOS Press
II. **Mathematics, Physics and Chemistry**	Kluwer Academic Publishers
III. **Computer and Systems Science**	IOS Press
IV. **Earth and Environmental Sciences**	Kluwer Academic Publishers
V. **Science and Technology Policy**	IOS Press

The NATO Science Series continues the series of books published formerly as the NATO ASI Series.

The NATO Science Programme offers support for collaboration in civil science between scientists of countries of the Euro-Atlantic Partnership Council. The types of scientific meeting generally supported are "Advanced Study Institutes" and "Advanced Research Workshops", although other types of meeting are supported from time to time. The NATO Science Series collects together the results of these meetings. The meetings are co-organized bij scientists from NATO countries and scientists from NATO's Partner countries – countries of the CIS and Central and Eastern Europe.

Advanced Study Institutes are high-level tutorial courses offering in-depth study of latest advances in a field.
Advanced Research Workshops are expert meetings aimed at critical assessment of a field, and identification of directions for future action.

As a consequence of the restructuring of the NATO Science Programme in 1999, the NATO Science Series has been re-organised and there are currently Five Sub-series as noted above. Please consult the following web sites for information on previous volumes published in the Series, as well as details of earlier Sub-series.

http://www.nato.int/science
http://www.wkap.nl
http://www.iospress.nl
http://www.wtv-books.de/nato-pco.htm

Series II: Mathematics, Physics and Chemistry – Vol. 111

Spin Structure of the Nucleon

edited by

Erhard Steffens
Physics Institute,
University of Erlangen-Nürnberg, Erlangen, Germany

and

Revaz Shanidze
Physics Institute,
University of Erlangen-Nürnberg, Erlangen, Germany
Permanent address:
High Energy Physics Institute,
Tbilisi State University, Georgia

Kluwer Academic Publishers

Dordrecht / Boston / London

Published in cooperation with NATO Scientific Affairs Division

Proceedings of the NATO Advanced Research Workshop on
Spin Structure of the Nucleon
Nor-Amberd, Yerevan, Armenia
29 June–4 July 2002

A C.I.P. Catalogue record for this book is available from the Library of Congress.

ISBN 1-4020-1440-6 (HB)
ISBN 1-4020-1441-4 (PB)

Published by Kluwer Academic Publishers,
P.O. Box 17, 3300 AA Dordrecht, The Netherlands.

Sold and distributed in North, Central and South America
by Kluwer Academic Publishers,
101 Philip Drive, Norwell, MA 02061, U.S.A.

In all other countries, sold and distributed
by Kluwer Academic Publishers,
P.O. Box 322, 3300 AH Dordrecht, The Netherlands.

Printed on acid-free paper

Table of Contents

Preface

The NATO Advanced Research Workshop on ,Spin Structure of the Nucleon' was held from June 29 to July 4, 2002, at the conference center 'Nor-Amberd' of the Yerevan Physics Institute (YerPhI) on a pitoresque slope of Mount Aragats, 2000m above sea level. The solitary location within a beautiful nature contributed greatly to the intense focus on the workshop subject. One afternoon and the last day was dedicated to explore some of the rich cultural and historical sites of Armenia.

This workshop has been the third of its kind. The first was held in July 9-11, 2001, Zeuthen, Germany (Topical Workshop on Transverse Spin Physics), and the second was held in August 2-7, 2001, Dubna, Russia (9th International Workshop on High-Energy Spin Physics, SPIN 01). The lively discussions and exchanges of ideas among participants at the NATO-ARW workshop in Nor-Amberd indicate that the subject is of great interest and future workshops in a time interval of two to three years should be considered.

The purpose of the workshop was to bring together the community from NATO, CIS and other countries working in this field, to discuss the state of the art of experiments and theory, and to establish a consensus on future scientific goals. There were 25 participants from NATO countries, 18 from the CIS, and 3 from elsewhere. About 30 original reports were presented, dealing with the exploration of hadronic spin structure via lepton scattering and with advancements in theoretical investigations of hadron spin physics from pertubative and non-perturbative QCD models, helping to strengthen the links between these two fields of activities.

The program covered the following topics: reviews of deep inelastic lepton scattering data; presentations of the most recent experimental results from DESY, CERN, SLAC and JLAB; the RHIC spin program at STAR; proton and neutron spin structure functions and the flavor structure of the nucleon; polarization observable in N-Delta transition; quark distributions in polarized ρ-mesons and its comparison with pions; deeply virtual Compton scattering (DVCS); theory of skewed parton distribution (SPD); spin effects in vector meson and exclusive meson production; polarized strangeness in semi-inclusive DIS; single spin azimuthal asymmetries; single spin asymmetry in heavy quark photoproduction and decays; the Q^2 dependence of GDH integral at HERMES; measurement of the GDH sum rule at SLAC; nucleon electromagnetic form factors; hyperon electroproduction; theoretical model for Lambda polarization in DIS; transverse polarization of Lambda and Anti-Lambda produced inclusively in eN-scattering; review of the nuclear attenuation effect: theory and experiment; YerPhI-TJNAF Collaboration: status and perspectives; study of nucleon correlation in nuclei; CLAS new data; vector meson production in the nuclear Coulomb field; QED effects in heavy ions collisions; prospects of Electron-Ion-Collider (EIC); activities towards a high-luminosity fixed target eN-experiment in Europe; the transverse polarized target for the HERMES experiment; and Compton polarimeters for TESLA and HERA.

The presentation of current experimental and theoretical studies led to new ideas and future plans. Today our attention of spin physics is focused on the detailed contributions of quarks and gluons to the nucleon spin, on theoretical and experimental investigations of generalized parton distributions and on the transverse spin structure, incl. the test of fundametal sum rules. There was a consensus among the workshop participants, that at present a transition takes place from inclusive spin physics to semi-inclusive and exclusive reactions, which require high polarization of beam and target and a new quality of hadron detection in lepton-nucleon scattering experiments. For this purpose, a new fixed-target facility with a high duty cycle providing polarized beams in the energy range 25-100 GeV could play a central role.

The Workshop programme was set up by N. Akopov and R. Avakian (YerPhI), D. Ryckbosch (Gent) and M. Amarian (DESY). The local organization was led by A. Avetisyan, supported by a strong team of YerPhI staff, including the workshop secretary Ivetta Keropyan. They made our stay very enjoyable and provided a warm atmosphere throughout the whole workshop. Particularly unforgettable was the workshop dinner with a large number of speeches and songs presented by the participants which gave us an impressive insight into armenian hospitality and style of living. I would like to thank the Yerevan Physics Institute for the hospitality and care extended to the participants.

The workshop has been supported very generously by the NATO Science Affairs Division, and my thanks are due to Dr. F. Pedrazzini, the director of the programme. We are also very grateful to the support provided by DESY (Hamburg) and YerPhI (Yerevan).

I am especially indebted to my co-director of the workshop, Academician Robert Avagian for making this event possible, and to his wife Lena for giving the participants and spouses a deep, impressive flavor of the armenian history and culture. Finally my thanks are due to all the speakers, the session chairs and the participants who contributed so much to the grand success of the workshop.

(Erhard Steffens, workshop co-director)

PROGRAMME OF THE NATO ADVANCED RESEARCH WORKSHOP
"SPIN STRUCTURE OF THE NUCLEON"
June 30 – July 3, 2002, Nor-Amberd, Yerevan, Armenia

June 30

9.00-10.00	BREAKFAST
10.00-10.10	WELCOME - R.O. Avagian
10.10-10.50	P. Bosted: Using Spin to Probe Nucleon Structure at SLAC
10.50-11.30	H. Jackson: The Flavor Structure of the Nucleon as Revealed at HERMES
11.30-11.40	COFFEE BREAK
11.40-12.20	A. Freund: Demystifying Generalized Parton Distributions – an Introduction
12.20-13.00	W.D. Nowak: Activities Towards a High-Luminosity Fixed-Target eN-Experiment in Europe
13.00	LUNCH
17:45	COFFEE
18.00-18.30	A.G. Oganesian: Quark Distribution in Polarized ρ Meson and its Comparison with Pion
18.30-19.15	James J. Kelly: Nucleon Electromagnetic Form-Factors
19.15-20.00	A. Kotzinian: Hadron Production and Lambda Polarization in DIS
20.00	DINNER

July 1

9.00-10.00	BREAKFAST
10.00-10.40	M. Amarian: Deeply Virtual Compton Scattering and Exclusive Meson Production at HERMES
10.40-11.20	H. Avagian: Spin and Azimuthal Asymmetries in DIS
11.20-11.40	COFFEE BREAK
11.40-12.20	N. Bianchi: Nuclear Attenuation in DIS
12.20-13.00	V. Ghazhikhanian: The RHIC Spin Program at STAR
13.00	LUNCH
17.45	COFFEE
18.00-18.30	N. Akopov: The Q^2 Dependence of Gerasimov-Drell-Hearn Integral at HERMES
18.30-19.10	N. Ivanov: Single Spin Asymmetry in Heavy Quark Photoproduction and Decays
19.10-19.40	V. Gharibyan: Compton Polarimeters for TESLA
19.40-20.00	D. Reggiani: The Transverse Polarized Target for the HERMES Experiment
20.00	BANQUET

July 2

9.00-10.00	BREAKFAST
10.00-10.40	D. Ryckbosch: Vector Meson Production
10.40-11.20	O. Grebenyuk: Transverse Polarization of Lambda and Anti-Lambda Produced Inclusively in ⟨
	Scattering at HERMES
11.20-11.30	COFFEE BREAK
11.30-12.00	G. Mallot: Status and Prospects of the COMPASS Experiment at CERN
12.00-12.30	G. Aghuzumtsyan: CHARM Production at HERA Using the ZEUS Detector
12.30-13.00	A. Borissov: Spin Physics of Exclusive ρ^0 Production
13.00	LUNCH
17.45	COFFEE
18.00-18.30	A. Bruell: Prospects on Electron-Ion-Collider (EIC)
18.30-19.00	K. Egiyan: YerPhI-TJNAF Collaboration: Status and Perspectives.
19.00-19.30	K. Egiyan: Study of Nucleon Correlations in Nuclei, CLAS New Data.
19.30-20.00	H. Mkrtchyan: Measurements of Electron Electric Form-factor Using
	Polarized e-beam at JLAB (Hall C)
20.00	DINNER

July 3

9.00-10.00	BREAKFAST
10.00-10.40	S. Frullani: Polarization Observables in N-Delta Transition Induced by Polarized Electrons
10.40-11.20	A. Airapetyan: Coherent ρ Meson Production at HERMES
11.20-11.30	COFFEE BREAK
11.30-12.00	S. Gevorkyan: Vector Meson Production in the Nuclear Coulomb Field
12.00-12.30	E. Kuraev: QED Effects in Heavy Ion Collisions
12.30-13.00	D. Ryckbosch: Concluding Remarks
13.00	LUNCH
14:00-20:00	EXCURSION
20:00	DINNER

Poster:

M. Dalton: Polarized Photon Production and Polarimetry of High Energy Photons Using Diamond Crystals

Co-director of the NATO
workshop, Robert Avagian
(YerPhI), during the opening
session of the workshop

Participants in front of the Conference Center Nor-Amberd
at the slope of Mount Aragats near Yerevan (Armenia)

Visit to the castle of Amberd (VIIth century AD) and the nearby chapel
located about 10km from the Conference Center Nor-Amberd

Excursion to Etchmiadzin, the religious center of Armenia
and holy seat of the Apostolic Church since 301 AD

SPIN STRUCTURE OF THE NUCLEON AT SLAC

Recent Results and Future Perspectives

P.E. BOSTED
University of Massachusetts
Amherst, Massachusetts 01003 USA

Abstract. Recent measurements are reviewed of the spin structure functions g_2^p and g_2^d measured at SLAC over the kinematic range $0.02 \leq x \leq 0.8$ and $0.6 \leq Q^2 \leq 20$ $(\text{GeV/c})^2$, made by scattering 29.1 and 32.3 GeV longitudinally polarized electrons from transversely polarized NH_3 and ^6LiD targets. The future SLAC program includes an experiment (E161) to measure $\Delta G(x)$ through polarized open charm photoproduction. Dominance of the photon-gluon fusion process can be tested using linearly polarized photons. Another experiment (E159) will determine the high-energy convergence of the GDH sum rule for both proton and deuteron target.

1. Results on g_1

The final results from SLAC E155 have now been published [1]. The structure functions g_1^p and g_1^n were measured in a single experimental setup over the large kinematic range $0.014 < x < 0.9$ and $1 < Q^2 < 40$ GeV^2 using deep-inelastic scattering of 48 GeV longitudinally polarized electrons from polarized protons and deuterons. The higher beam energy of E155 allowed a significant extension of the kinematic range of the earlier E143 experiment. The data indicate that the Q^2 dependence of g_1^p (g_1^n) at fixed x is very similar to that of the spin-averaged structure function F_1^p (F_1^n). Simple empirical fits to the data are given by

$$\frac{g_1^p}{F_1^p} = x^{0.700}(0.817 + 1.014x - 1.489x^2)(1 - \frac{0.04}{Q^2}) \qquad (1)$$

$$\frac{g_1^n}{F_1^n} = x^{-0.335}(-0.013 - 0.330x + 0.761x^2)(1 + \frac{0.13}{Q^2}). \qquad (2)$$

From an NLO QCD fit to all available data, E155 finds that the difference of first moments $\Gamma_1^p - \Gamma_1^n = 0.176 \pm 0.003 \pm 0.007$ at $Q^2 = 5$ GeV^2, in

1

E. Steffens and R. Shanidze (eds.), Spin Structure of the Nucleon, 1–8.

agreement with the Bjorken sum rule prediction of 0.182 ± 0.005. Using the same NLO pQCD fit, the quark singlet contribution in the $\bar{M}S$ scheme is $\Delta\Sigma = 0.23 \pm 0.04(\text{stat}) \pm 0.06(\text{syst})$ at $Q^2 = 5$ GeV^2, confirming earlier indications that quarks carry only a small fraction of the spin of the nucleon. While theoretical errors are large, the data clearly favor a positive sign for $\Delta G(x)$.

2. The g_2 structure function

2.1. THE EXPERIMENT

The recent (1999) experiment SLAC E155x made the best measurements of g_2 in the DIS region for the proton and deuteron to date. The final results have recently been submitted for be publication [2]. The experiment used the 120 Hz SLAC electron beam with a longitudinal polarization of $(83 \pm 3)\%$ at energies of 29.1 and 32.3 GeV and a typical current of 25 nA. Transversely polarized NH$_3$ and ^{6}LiD targets were used as sources of polarized protons (average polarization 75%) and deuterons (average polarization 20%). Scattered electrons were detected in three independent spectrometers centered at $2.75°$, $5.5°$, and $10.5°$. Electrons in each spectrometer were separated from pions using gas Cherenkov counters and segmented electromagnetic calorimeters. Tracking was done with scintillator hodoscopes.

2.2. CORRECTIONS

The physics asymmetry $A(x, Q^2)$ was determined according to

$$A(x,Q^2) = \frac{R^{\uparrow\downarrow} - R^{\uparrow\uparrow}}{R^{\uparrow\downarrow} + R^{\uparrow\uparrow}} \frac{1}{C_1 P_B P_T f} + C_2 A_p(x,Q^2) \frac{\sigma_p}{\sigma_d} \tag{3}$$

where $P_B P_T f$ accounts for beam polarization, target polarization, and dilution factor. [Radiative corrections are also made]. The nuclear corrections are contained in the C_1 and C_2 terms, which depend only slightly on x and Q^2, and take into account scattering from nuclei other than free proton or deuterons in the solid polarized targets. For NH$_3$, $C_2 = 0$ by definition, and C_1 accounts for polarized ^{15}N, which is polarized opposite to free protons because it acts like single proton "hole" [3]. Numerically, $C_1 \approx 1 - 0.11 * P_N/P_p$ ranges from 1.01 to 1.04, where P_N (P_p) is the nitrogen (proton) polarization. The ratio P_N/P_p was measured as a function of P_p with a special NMR setup, and a fit was used to determine the C_1 correction for the varying values of P_p during the main E155x data taking. For LiD $C_1 \approx 1.86$ because the nuclear wave function of ^{6}Li is similar to 0.86 of a free polarized deuteron, plus a spectator unpolarized α particle [3]. The

LiD material used in E155x contained 4% of the ^{7}Li isotope, which has an unpaired proton, and gives a non-negligible C_2 correction, which multiplies the measured proton asymmetry at the same (x, Q^2) at which the deuteron asymmetry is obtained. Typically, C_2 was -0.042 for E155x.

2.3. RESULTS

Since the results for g_2 in the three spectrometers and two beam energies are reasonably consistent with the Q^2 dependence of the twist-two g_2^{WW} model, they are averaged together using this assumption to produce the averaged values shown in Fig. 1.

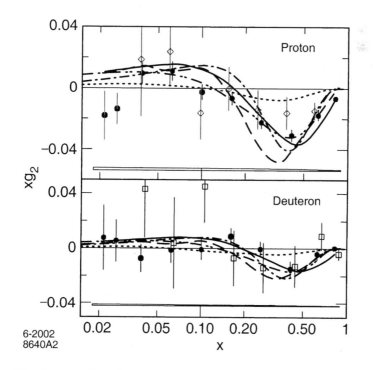

Figure 1.   The structure function xg_2 averaged over the three spectrometers for E155x (solid circles), and data from E143 [4] (diamonds) and E155 [5] (squares). The errors are statistical; the systematic errors are shown at the bottom of each panel. Also shown is g_2^{WW} at the average Q^2 of this experiment at each value of x (solid curves) and the calculations of Stratmann [6] (dash-dot-dot), Gamberg and Weigel [7] (dash-dot), Song [8] (dot), and Wakamatsu [9] (dash).

The proton results are clearly different than zero, and exhibit an x-dependence similar to that of the g_2^{WW} model. There appear to be statistically significant differences from the g_2^{WW} model, possibly indicating non-zero twist-3 contributions. The data are in qualitative agreement with the bag model calculation of Stratmann [6] and the chiral soliton calculation of Gamberg and Weigel [7], but are considerably more negative than the model of Song [8]. The deuteron data have larger errors than the proton data, but also indicate significantly negative values at high x, and are in qualitative agreements with g_2^{WW}, Stratmann [6], and Gamberg and Weigel [7].

3. Future: $\Delta\sigma^{\gamma N}(k)$ and the High Energy Contribution to the GDH Sum Rule

An experiment (E159 [10]) has recently been approved at SLAC to measure $\Delta\sigma^{\gamma N}(k)$, the helicity-dependent total photo-absorption cross section, for photon energies $5 < k < 40$ GeV, on both proton and deuteron targets. The first goal is to complement the extensive set of measurements of g_1 at $Q^2 > 0$ with the anchor points at $Q^2 = 0$, useful for global fitting [11, 12] and understanding the low-x behavior. The second goal is to test the convergence of the GDH sum rule [13],

$$\int_{k_\pi}^{\infty} \frac{dk}{k} \Delta\sigma^{\gamma N}(k) = \frac{2\pi^2 \alpha \kappa^2}{M^2} \tag{4}$$

where M and κ are the nucleon mass and anomalous magnetic moment, and k_π is the threshold energy needed to produce at least one pion. Early indications from measurements in the resonance region are that the sum rule may already be over-saturated, requiring a sign change to $\Delta\sigma^{\gamma N}(k)$ for convergence.

The experiment will use an untagged coherent bremsstrahlung beam to create a high flux of circularly polarized photons. With coherent bremsstrahlung, a set of high intensity spikes is generated by proper orientation of a diamond crystal radiator. With longitudinally polarized electrons, the incoherent bremsstrahlung photons are circularly polarized, with the polarization maximal at the endpoint. The coherent photons are elliptically polarized: the circular component is almost identical to that for incoherent photons. The coherent peak polarization also has a linear component which will cancel in the measurement of $\Delta\sigma^{\gamma N}(k)$, but will allow for the measurement of possibly interesting azimuthal asymmetries.

For targets, we will use polarized NH_3 and ND_3 as sources of polarized protons and neutrons. Polarized deuterons,to first order, allow measurements of the isovector combination $(n+p)/2$, with small corrections for the deuteron D-state, shadowing, and nuclear coherent hadron production. An extension to this proposal could use a polarized ^{3}He target to verify the consistency of $\Delta\sigma^{\gamma n}(k)$ for the neutron as extracted from either deuterium or ^{3}He. The detector is a simple calorimeter optimized to measure $> 98\%$ of all hadronic interactions, and to reject electromagnetic backgrounds.

The expected errors are shown in Fig. 2 for both the proton and neutron, and for two data taking modes, one involving counting each hadronic interaction individually, and one where only the total flux of hadrons for each helicity state is measured. Even with the larger counting mode statistical errors, a very good determination can be made of both the magnitude and energy dependence of $\Delta\sigma^{\gamma N}(k)$ for $5 < k < 40$ GeV. By measuring with both proton and deuteron targets, the high energy contributions to both the isovector and isoscalar GDH sum rules can be determined. This will allow tests of Regge-inspired models, which predict very different behavior for the isovector and isoscalar contributions, and will provide a baseline for studies of the polarized spin-structure functions measured with virtual photons. Ultimately, the scale of convergence of the GDH sum rule is a measure of the energy scale at which spin excitations of the nucleon are important.

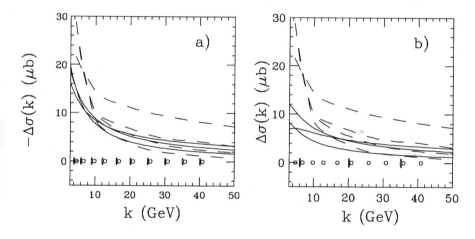

Figure 2.   a) Projected proton error bars for $\Delta\sigma^{\gamma p}(k)$ for E159 as a function of photon energy for the counting mode (rectangles) and flux integration mode (circles). The dashed curves are representative models from Ref. [12], the solid curves are from Ref. [11]; b) same but for the neutron as measured with ND_3.

4. Future: Polarized Charm Photoproduction and the Gluon Spin

Another approved SLAC experiment using the circularly polarized photon facility mentioned above is E161 [14], designed to study the gluon spin structure of nucleons using open charm photoproduction. The measurements will utilize a ^{6}LiD polarized target to measure the asymmetry A_{cc} in open charm photoproduction. This process is dominated by the photon-gluon fusion mechanism. The open charm signal will be measured by detecting the muons from charm decay at large p_T. This experiment will measure the asymmetry A_{cc} over a range of energies and p_T sensitive to x from 0.1 to 0.3 with statistical precision of about 0.01. This is to be compared with the range of current theoretical models in which the values of A_{cc} differ by more than 0.1 and $x\Delta g(x)$ differ by up to 0.3 in this x range.

Figure 3 shows the expected statistical error on A_{cc} as as function of p_T^μ for $5 < P_\mu < 10$ GeV for three incident photon coherent peak energies. The points are arbitrarily plotted at a value of zero. Also shown are the calculated asymmetries from a sample of gluon polarization models. The systematic errors of 10% of the value of the asymmetry (typical error 0.01) will be highly correlated point-to-point. There will be additional data for higher momentum muons. The statistical errors are projected to be smaller than for the similar COMPASS experiment [15] currently running at CERN, but the lower photon energies of E161 correspond to larger values of x for the gluons.

The E161 experiment will also measure the double spin asymmetry for elastic and inelastic photoproduction of closed charm (J/ψ particles). The latter may also yield interesting information on the gluon spin, if the relative contributions from color singlet and color octet mechanisms can be reliably modeled.

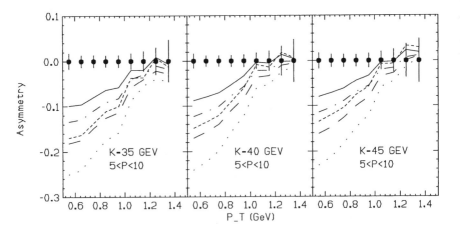

Figure 3. The E161 projected statistical errors on the asymmetry A_{cc} for open charm production as a function of p_T^μ of the detected muon for $5 < P_\mu < 10$ GeV. Also shown are asymmetries for several gluon polarization models.

5. Future: Linear Polarization Asymmetry in Charm Photoproduction

A third planned experiment at SLAC, E160 [16], will use unpolarized electrons to make coherent bremsstrahlung beams at 15, 25, and 35 GeV. These photons will have a fairly high degree of linear polarization. While the main goal of the experiment is to measure the A-dependence of J/ψ and ψ' quasi-elastic photoproduction, we will also measure the linear polarization single-spin asymmetry for nuclear coherent, quasi-elastic, and inelastic J/ψ photoproduction "for free". At present, I do not know of any predictions for these asymmetries. However, there is a QCD prediction [17] for open charm photoproduction at SLAC photon energies: the single-spin asymmetry is predicted to be large, at about 0.2, and unlike the cross section itself, is quite stable against higher order QCD corrections. Open charm events will be tagged in the spectrometer by single muons with transverse momenta near 1 GeV, where the backgrounds from π, K, and J/ψ decays are the smallest. Recent calculations [18] show that for muons with $p_T > 1$ GeV/c, the asymmetry is practically the same as for the charmed quarks themselves, or order 0.2. Preliminary estimates are that the asymmetry can be measured with a statistical error of about 0.02 or better, which would provide a meaningful test of the prediction. The QCD predictions are practically independent of muon energy (for $p_T > 1$ GeV), and are the same for positive and negative muons, assuming that only the photon-gluon fusion diagram contributes. By making measurements for both sign muons, as a function of muon energy, the experiment will be able to determine to what extent soft processes, such as diffractive Pomeron exchange, or associated production of Λ_c are important contributors to the total open charm cross section. This will be of great value in the the interpretation of the results with circularly polarized photons and polarized target, from which E161 (and the COMPASS experiment at CERN) wish to extract information of $\Delta G(x)$.

6. Summary

The recent SLAC data on g_1 and g_2 have provided significant new information of the spin structure of the nucleon. Future experiments using polarized photon beams should provide insight into the role of gluon polarization in the nucleon, and the behavior of the spin structure functions in the limit of $Q^2 = 0$.

This work was supported by the National Science Foundation and the Department of Energy contract DE–AC03–76SF00515.

References

1. E155 Collaboration, Phys. Lett. B463 (1999) 339; Phys. Lett. B493 (1999) 19.
2. E155x Collaboration, SLAC–PUB–8813, hep-ex/0204028, submitted to Phys. Rev. Lett. (2002).
3. S. Bueltmann et al., Nucl. Instrum. Meth. A 425 (1999) 23.
4. E143 Collaboration, Phys. Rev. Lett. 76 (1996) 587; Phys. Rev. D 58 (1998) 112003.

5. E155 collaboration, Phys. Lett. B 458 (1999) 529.
6. M. Stratmann, Z. Phys. C 60 (1993) 763.
7. H. Weigel, L. Gamberg, and H. Reinhart, Phys. Rev. D 55 (1997) 6910.
8. X. Song, Phys. Rev. D 54 (1996) 1955.
9. M. Wakamatsu, Phys. Lett. B 487, 118 (2000).
10. http://www.slac.stanford.edu/exp/e159
11. N. Bianchi, E. Thomas, Phys. Lett. B 450 (1999) 439; E. Thomas, N. Bianchi, Nucl. Phys. Proc. Suppl. 82 (2000) 256.
12. S.D. Bass and M.M. Brisudova, Eur. Phys. J. A4 (1999) 251; S.D. Bass, Mod. Phys. Lett. A12 (1997) 1051 and references therein.
13. S. D. Drell and A. C. Hearn, Phys. Rev. Lett 16, 908 (1966); S. B. Gerasimov, Yad. Fiz. 2, 598 (1966); S.J. Brodsky and J.R. Primack, Ann. Phys. 52 (1969) 315.
14. http://www.slac.stanford.edu/exp/e161
15. COMPASS proposal, CERN/SPSLC-96-14 (March, 1996).
16. http://www.slac.stanford.edu/exp/e160
17. N.Ya. Ivanov, A. Capella and A.B. Kaidalov, Nucl. Phys. B586 (2000), 382 and N.Ya. Ivanov, Nucl. Phys. B615 (2001), 266.
18. N.Ya. Ivanov, P. Bosted, K. A. Griffioen, manuscript in preparation.

FLAVOR STRUCTURE OF THE NUCLEON
AS REVEALED AT HERMES

H.E. JACKSON
Physics Division, Argonne National Laboratory
Argonne, Illinois 60439, USA
(On behalf of the HERMES Collaboration)

Abstract. The flavor structure of the nucleon as revealed in parton distributions (PDF's) is central to understanding the partonic structure of the nucleon. Recent data on unpolarized PDF's and their implications for the flavor- dependent quark helicity distributions are discussed. Results are presented for spin asymmetries in inclusive and semi-inclusive cross sections for production of pions, and kaons measured by the HERMES experiment at DESY in deep-inelastic scattering of polarized positrons on proton and deuterium targets. A full 5 component extraction of polarized quark distributions for u, d, $\overline{u}$, $\overline{d}$, and $(s + \overline{s})$ is reported. Resulting valence quark distributions conform to results of earlier experiments. There is no evidence for a significant polarization of the light sea. In contrast to the conclusions inferred from studies of polarized inclusive scattering, a leading order analysis of the HERMES data suggests a zero or slightly positive polarization of the strange sea. There is no evidence for a measurable flavor asymmetry in the helicity distributions for the light sea.

1. Introduction

Parton distribution functions(PDF's) form the basis for the description of the flavor structure of the nucleon. The unpolarized parton distribution functions, $q_f(x,Q^2)$, where f is the quark flavor, describe the momentum structure of the nucleon. Polarized parton distribution functions, such as the longitudinal PDF, $\Delta q_f(x,Q^2)$, provide information on the helicity distributions of the quarks, i.e. the spin structure of the nucleon. Over the years a very detailed picture of unpolarized PDF's has emerged, but a number of questions remain, some with important implications for the spin

E. Steffens and R. Shanidze (eds.), Spin Structure of the Nucleon, 9–20.

structure of the nucleon. By contrast, for polarized PDF's, only broad features of parton helicity distributions are established. Detailed properties, such as the sea polarizations are poorly known. And there are no direct measurements of gluon polarization, or of contributions from parton orbital angular momenta. New experiments addressing these issues involving selective probes of various features of partonic structure, particularly spin structure, are in progress or about to be launched at HERMES, COMPASS, and RHIC. This paper focuses on recent progress in measurements of the flavor dependence of the helicity distributions of the nucleon as measured in the HERMES experiment. The discussion begins with brief remarks on unpolarized PDF's and a discussion of models of PDF's and their implications for deep inelastic scattering(DIS) reactions, and in particular the flavor structure of polarized PDF's. This is followed by a description of the HERMES experiment and the flavor tagging technique for isolating quark effects in parton distributions. Finally, new data on polarized quark helicity distributions and their implications are discussed.

2. Unpolarized parton distributions

A new generation of parton distributions based on a global data analysis (CTEQ6) is now available [1]. The study is carried out mainly in the $\overline{MS}$ scheme with an improved error treatment, but in most respects the new PDF's do not differ substantially from the previous compilation, CTEQ5. Among the outstanding problems are lack of data on parton distributions at high x_{bj}, uncertainties in the behaviour of the light sea flavor asymmetry for x_{bj} above 0.1, and the fragmentary information on the distributions for strange quarks. The unresolved discrepency in s(x) as determined by comparing dimuon ν data to results from charged and neutral current structure functions is particularly serious. The lack of knowledge of s(x) precludes a precise comparision of new results for Δs(x) with the unitary limit. The CTEQ compilation assumes $s = \bar{s} = 0.2(\bar{u} + \bar{d})$. The decrepency between the CTEQ value for s(x) and other compilations is as much as 100%.

The flavor asymmetry of the light sea has received much attention. The Gottfried sum rule which is given by the equation

$$I_G = \int_0^1 [F_2^p(x) - F_2^n(x)]\frac{dx}{x} = \frac{1}{3} - \frac{2}{3}\int_0^1 [\bar{d}_p(x) - \bar{u}_p(x)]dx \qquad (1)$$

provides a measure of that asymmetry. The first evidence for an asymmetry came from the CERN experiment NA51, which reported the result, $I_G = 0.235 \pm 0.026$ [2], instead of the value 1/3 expected for a symmetric sea. Subsequently, in a measurement of the Drell Yan process [3] the ratio, $\bar{d}/\bar{u}$ was measured as a function of x. The results are presented in fig. (1).

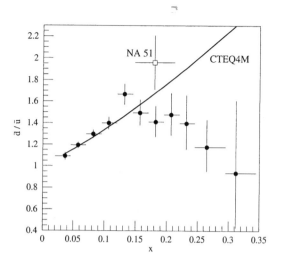

Figure 1.   The ratio of $\bar{d}/\bar{u}$ as a function of the scaling variable x measured in the Fermilab experiment E866. The result from NA51 is plotted as an open box. Also shown is the CTEQ4 prediction(solid line).

The flavor-asymmetry of the light-quark sea was confirmed by the HER-MES experiment [4] using an entirely different reaction, DIS. Important features of these data can be explained by a number of theoretical models, developed to describe PDF's. According to these models, the flavor aysm-metry observed for the light sea puts important constraints on the flavor asymmetry for the polarization of the light sea.

The chiral quark soliton model(χQSM) makes a strong prediction [5] about $\Delta\bar{u} - \Delta\bar{d}$. In this approach the nucleon is described as a soliton of an effective field theory. Unpolarized quark and antiquark PDF's, as well as polarized PDF's for quarks plus antiquarks given by this model are in good agreement with experiment. From the structure of the model, para-metrically $\Delta\bar{u}(x) - \Delta\bar{d}(x) > |\bar{u}(x) - \bar{d}(x)|$. Calculations at low scale in the large-N_c limit with an effective field theoretic description of chiral symme-try breaking confirm this prediction. The results are shown in Fig. (2). A second very different model employs methods of statistical mechanics. This statistical model in which one generates all the PDF's from a very small number of parameters [6] shows impressive agreement with the global data base for the structure functions $F_2^p(x, Q^2), F_2^d(x, Q^2)$, and, $G(x, Q^2)$. Again, from the chiral structure of QCD and features of u(x) and d(x) known from DIS, in this theory, $\bar{d}(x) > \bar{u}(x)$; $\Delta\bar{u}(x) > 0$; $\Delta\bar{d}(x) < 0$, and most signif-icantly, $\Delta\bar{u}(x) - \Delta\bar{d}(x) \sim \bar{d}(x) - \bar{u}(x)$. A third model developed to explain the enhancement of $\bar{d}$ over $\bar{u}$ is a meson cloud model [7] in which the phys-ical proton is viewed as a coherent sum of virtual meson-baryon states.

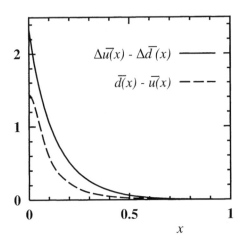

Figure 2. Isovector antiquark distributions at low normalization point($\mu = 600$ MeV), as obtained from the chiral quark soliton model.

The leading term is a proton with a symmetric sea. The "Sullivan process" generates an antiquark asymmetry through reactions which arise from the πNN and the $\pi\Delta$N couplings. With some tuning of the cutoff parameters for these couplings, the virtual pion model gives an excellent description of the measured light sea asymmetry. However, in contrast to other models, for a light sea generated by a pion cloud plus a symmetric sea generated by gluon splitting, this model gives $\Delta\bar{u} - \Delta\bar{d} \approx 0$. Because of these differing predictions for the light sea polarization, a precise measurement of the asymmetry in the helicity distributions of the light sea will provide a good test of these models. A detailed measurement of the flavor structure of the quark helicity distributions is one of the principal goals of the HERMES experiment.

3. The HERMES Experiment

Deep inelastic scattering events are generated in the HERMES experiment by the interaction of the polarized lepton beam of the HERA accelerator at DESY with polarized target gases which are injected into a 40 cm long, tubular open-ended storage cell located at an interaction point on the lepton orbit. The lepton beam is self polarized by the Sokolov-Ternov effect with a polarization time which is typically about 20 minutes. The beam polarization is measured continuously with Compton backscattering of circularly polarized laser beams. The beam polarization is routinely about 0.55. Spin rotators in the ring provide longitudinal polarization at the interaction point. The polarized target gases, atomic H or D, are generated

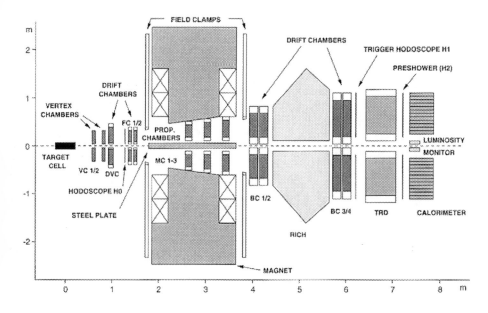

Figure 3.   The HERMES Spectrometer.

by an atomic-beam source based on Stern-Gerlach separation which pro-
vides an areal density of about 2×10^{14} atoms/cm^2 for H and 7×10^{13}
atoms/cm^2 for D. The nuclear polarization is measured with a Breit-Rabi
polarimeter and the atomic fraction with a target gas analyser. The target
polarization is reversed within short time intervals to minimize systematic
effects. The relative luminosity is measured by detecting Bhabha- or Mott-
scattered target electrons in coincidence with the scattered lepton, in a pair
of NaBi(WO$_4$)$_2$ electromagnetic calorimeters.

The HERMES spectrometer[8], shown in Fig. (3), is a forward angle
open geometry system consisting of two halves which are symmetric about
a central horizontal shielding plate in the spectrometer magnet. A fly's eye
calorimeter and a transition radiation detector furnish clean separation of
hadrons and leptons. Identification of $\pi's, K's$ and $p's$ is accomplished by
means of a novel dual-radiator ring-imaging Cerenkov counter(RICH)[9],
which is located between the rear tracking chambers. The combination of
radiators consisting of a wall of clear aerogel and a gas volume of C_4F_{10}
provide clean particle identification over almost the full acceptance of HER-
MES, i.e. 2-15 GeV. The scattered leptons and hadrons produced within
an angular acceptance of $\pm$ 170 mr horizontally, and 40 - 140 mrad verti-
cally are detected and identified. Typical kinematics for studies of DIS are
$E = 27.5$ GeV for the incident lepton, $x > 0.02$ where $x = Q^2/2M\nu$ is the
Bjorken scaling variable, 0.1 GeV$^2 < Q^2 < 15$ GeV^2 with $-Q^2$ the square

of the momentum transfer, and $\nu < 24$ GeV where $\nu = E - E'$ with $E(E')$ is the energy of the incoming(scattered) lepton in the target rest frame. To ensure that hadrons detected are in the current fragmentation region, cuts of $z = E_h/\nu > 0.2$ and $x_F \approx 2p^h_{parallel}/W > 0.1$ are imposed, where $W = \sqrt{2M\nu + M^2 - Q^2}$ is the invariant mass of the virtual photon-proton system.

4. Flavor Decomposition

Until recently, all of the information on the flavor decomposition of quark helicity distributions resulted from analyses of inclusive data under the assumption of SU(3) symmetry. One found [10]

$$\Delta u = +0.78 \pm 0.03, \quad \Delta d = -0.48 \pm 0.03, \quad \Delta s = -0.14 \pm 0.03. \quad (2)$$

A subsequent next to leading order global analysis of all data for a proton target [11] gives for the strange quark contribution to the singlet axial coupling, $a_s = -0.07 \pm 0.04$. In order to improve sensitivity to flavor dependencies, the HERMES experiment uses semi-inclusive deep-inelastic scattering to determine the separate contributions $\Delta q_f(x)$ of the quarks and antiquarks of flavor f to the total spin of the nucleon. By means of the technique of flavor tagging, individual spin contributions can be determined directly from spin asymmetries of hadrons with the appropriate flavor content. For example, the spin asymmetry of K^-, an all sea object, will have a high sensitivity to the polarization of the quark sea. The measured semi-inclusive spin asymmetry, $A^h_\parallel$, and the corresponding photon-nucleon asymmetry, A^h_1, for the hadron of type h are given by

$$A^{(h)}_\parallel = \frac{N^{\uparrow\downarrow}_{(h)} - N^{\uparrow\uparrow}_{(h)}}{N^{\uparrow\downarrow}_{(h)} + N^{\uparrow\uparrow}_{(h)}}, \quad A^h_1 = \frac{A^h_\parallel}{D(1 + \eta\gamma)}, \quad (3)$$

where, for simplicity, we assume unity beam and target polarizations and constant luminosity. Here D is the depolarization factor for the virtual photon, η and γ are kinematic factors, and $N^{\uparrow\uparrow}(N^{\uparrow\downarrow})$ are the number of DIS events with coincident hadrons for target polarization parallel (antiparallel) to the beam polarization. In leading order QCD assuming the validity of factorization, one can write the semi-inclusive DIS cross section, $\sigma^h(x, Q^2, z)$, to produce a hadron with energy E_h and energy fraction $z = E_h/\nu$ as

$$\sigma^h(x, Q^2, z) \propto \Sigma_f e_f^2 q_f(x, Q^2, z) D^h_f(x, Q^2) \quad (4)$$

where the sum is over quark and antiquark types $f = (u, \bar{u}, d, \bar{d}, s, \bar{s})$. $D^h_f(x, Q^2)$ is the fragmentation function for producing hadron h from a

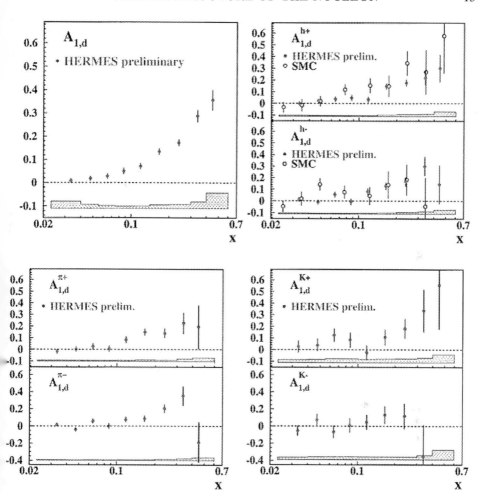

Figure 4. Inclusive and semi-inclusive hadron, pion and kaon asymmetries for a deuterium target. The hadron asymmetries are compared with data from the SMC collaboration. The error bars of the HERMES data are statistical and the bands are systematic uncertainties. These data form the deuteron portion of the data base used in the purity analysis described in the text.

quark of flavor f. The quark charge, e_f, is in units of the elementary charge. In this approximation,

$$A_1^h(x,z) = \frac{\int_{z_{min}}^1 dz \sum_f e_f^2 q_f(x) \cdot D_f^h(z)}{\int_{z_{min}}^1 dz \sum_{f'} e_{f'}^2 q_{f'}(x) \cdot D_{f'}^h(z)} \cdot \frac{\Delta q_f(x)}{q_f(x)} \cdot \frac{1 + R(x,Q^2)}{1+\gamma^2} \quad (5)$$

$$= \sum_f P_f^h(x) \frac{\Delta q_f(x)}{q_f(x)} \cdot \frac{1 + R(x,Q^2)}{1+\gamma^2}. \quad (6)$$

The quantities, $P_f^h(x)$, are the integrated purities which are defined by Eq. (6). They are spin-independent quantities in leading order and represent the probability that the quark, q_f was struck in the DIS event. The term containing the ratio $R = \sigma_L/\sigma_T$ of the longitudinal to transverse photon absorption cross section corrects for the longitudinal component that is included in experimental parameterizations of $q_f(x, Q^2)$ but not in $\Delta q_f(x, Q^2)$. The term $\gamma = \sqrt{Q^2}/\nu$ is a kinematic factor.

A Monte Carlo method based on a DIS physics event generator is used to calculate the purities from CTEQ5 leading order parameterizations of the parton distributions. A LUND string fragmentation model tuned to HERMES kinematics provides fragmentation probabilities. The Monte Carlo program includes the effects of the acceptance of the experiment. The systematic uncertainties in the purities were estimated by variation of the fragmentation parameters and by using alternate PDF parameterizations. By incorporating the correction factor in the purities, one can rewrite Eq. (6) in a matrix form as

$$\mathbf{A}(\mathbf{x}) = \mathbf{P}(\mathbf{x}) \cdot \mathbf{Q}(\mathbf{x}) \tag{7}$$

where $\mathbf{A}(\mathbf{x})$ becomes a vector whose elements are all the integrated measured asymmetries which are to be included in the analysis. The $\mathbf{Q}(\mathbf{x})$ vector contains the quark and antiquark polarizations. These quantities are now connected by the purity matrix which contains the effective integrated purities. The determination of the quark polarizations from the experimentally measured spin asymmetries is reduced to the task[12] of inversion of Eq. (7) to obtain $\mathbf{Q}(\mathbf{x})$. Eq. (7) can be solved for $\mathbf{Q}(\mathbf{x})$ by minimizing

$$\chi^2 = (\mathbf{A} - \mathbf{P} \cdot \mathbf{Q})^t V_A^{-1} (\mathbf{A} - \mathbf{P} \cdot \mathbf{Q}) \tag{8}$$

where V_A is the covariance matrix of the asymmetry vector $\mathbf{A}(\mathbf{x})$. The flavor decomposition is obtained by solving Eq. (7) for a vector $\mathbf{Q}(\mathbf{x})$ of a dimension corresponding to the number of independent quark flavor distributions.

5. Results

The purity formalism has been used in the HERMES analysis to make a flavor decomposition into polarized quark distributions for $u, \bar{u}, d, \bar{d}$, and $s + \bar{s}$. For the first time, a global analysis of inclusive spin asymmetries and semi-inclusive spin asymmetries for π^+, π^-, K^+, and K^- has been carried out for longitudinally polarized targets of hydrogen, and deuterium. The measured spin asymmetries $A_1^h(x, Q^2, z)$ were integrated in each x bin over the corresponding Q^2-range and the z-range from 0.2 to 0.8 to yield

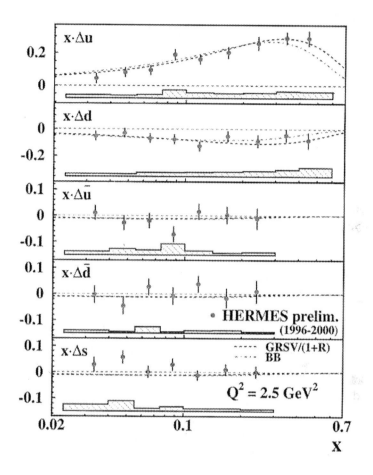

Figure 5. The x-weighted quark spin dependent densities. The plots show the results of a five parameter fit to the data assuming a symmetric strange sea polarization. The data have been evolved to a common $Q^2 = 2.5$ GeV^2. The dashed line shows a GRSV parameterization, and the dashed-dotted curve an alternate parameterization of Bluemlein and Boettcher (hep-ph/0203155).

$A_1^h(x)$. The data for the various semi-inclusive asymmetries together with the inclusive data obtained with the deuterium target are shown in Fig. (4). There the data for unidentified hadrons are seen to be in agreement with earlier results from the SMC collaboration [13].

The results of the decomposition obtained by solving Eq. (7) are presented in in Fig. (5). A symmetric strange sea polarization was assumed, i. e. $\Delta s/s = \Delta \bar{s}/\bar{s}$. The general features of the quark densities follow those of earlier decompositions[12, 13]. The u-quarks show a strong positive po-

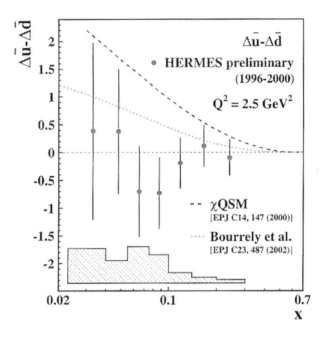

Figure 6.   Flavor asymmetry $\Delta u - \Delta d$ of the light sea extracted from the HERMES five-component purity analysis. The curves describe predictions of the χQSM model and the statistical model (see text). The error bars give statistical uncertainties and the shaded band the systematic error.

larization, while the d-quarks have a substantial negative polarization. The non-strange sea quarks are not significantly polarized. However, the strange sea appears to be positively polarized, contrary to the conclusions drawn within leading order QCD analysis[14] based on earlier inclusive data. The triplet strength $\Delta q_3 = \Delta u - \Delta d$ extracted from the HERMES data is in agreement with the Bjorken sum rule. The polarization of the strange sea can be extracted directly from the same data set by means of a purity analysis which uses only two spin asymmetries, $A_1^D(x)$ and $A_{1,D}^{K^+ + K^-}(x)$. For this analysis, fragmentation functions from e^+e^- collider experiments can be used to calculate purities. This method measures the quantity $\Delta s + \Delta \bar{s}$ with no assumption about the strange sea symmetry, and provides an independent check of the result from the five-component decomposition. The results obtained show the same trend of positive strange sea polarization.

6.   Remarks and conclusions

The HERMES result is the first "direct" five component decomposition of quark helicity distributions using flavor tagging. The distributions for u, d, and the nonstrange sea show trends similar to those inferred from inclusive measurements, i.e. $\Delta u >> 0$, $\Delta d < 0$, and $\Delta \bar{u} = \Delta \bar{d} \approx 0$. The result $\Delta(s + \bar{s}) > 0$ is unexpected, and it poses a challenge to our understanding of the quark sea. To the extent that the sea arises from gluon splitting, one would expect from the result for the strange sea that the light sea should show similar trends, i.e. $\Delta \bar{u} \approx \Delta \bar{d} > 0$. The precision of the result reported here is limited, but its implications are of such significance, that further measurements of improved accuracy should receive high priority. However, one point is clear. The hypothesis of a large negative strange sea polarization as an explanation for the apparent violation of the Ellis-Jaffe sum rule in inclusive scattering is ruled out.

The HERMES flavor decomposition does not support recent conjectures of a strong breaking of the flavor symmetry of the light sea. The results for the quantity $\Delta \bar{u} - \Delta \bar{d}$ are shown in Fig. (6) together with predictions based on the chiral quark soliton model[5] (χQSM) and the statistical model of Bourrely, et al [6]. Although the statistics are limited, the data indicate that any flavor asymmetry in the nonstrange sea is substantially smaller than the prediction of the χQSM, and each of the measured points lies below the prediction of the statistical model. At the same time the HERMES result is consistent with the expectation based on the simple pion-cloud model discussed in section 2, that the polarized asymmetry should be small. These results demonstrate that details of the flavor decomposition of the helicity distributions can play an important part along with those of the unpolarized distributions in providing insight into the mechanism by which the quark sea of the nucleon is generated.

Other details of the sea distributions are under study. The singlet strength $\Delta q_3 = \Delta u - \Delta d$, agrees to within 1σ with the Bjorken sum rule which states

$$\Delta q_3 = \int_0^1 \Delta q^{NS}(x)dx = |\frac{g_a}{g_v}| \times C_{QCD} = 1.01 \pm 0.05 \qquad (9)$$

A direct test of the SU(3) symmetry assumed in inclusive analyses carried out to date, can be made with the data from the flavor decomposition, by determining the octet strength $\Delta q_8 = \Delta u + \Delta d - 2\Delta s$. The octet combination can be related with SU(3) symmetry to the hyperon decay constants F and D according to the relation

$$\Delta q_8 = (3F - D) \times C_{QCD} = 0.46 \pm 0.03 \qquad (10)$$

where C_{QCD} is taken from ref. [15]. Analysis of the HERMES data is continuing. A determination of Δq_8 is of high priority. In summary, the HERMES results are the first complete decomposition of the quark helicity contributions to the spin of the nucleon. Such data will be of great value in revealing the manner in which the nucleon spin is generated by its partonic structure.

Acknowledgements

The support of the DESY management and staff and the staffs of the collaborating institutions is gratefully acknowledged. The author wishes to thank Gerard van der Steenhoven and Roy J. Holt for careful readings of the manuscript. The author also acknowledges the massive efforts of all the HERMES collaborators which have made the program a success. This work is supported in part by the U.S. Department of Energy under Contract No. W-31-109 -ENG-38.

References

1. Pumplin, J., *et al.* (2002) New Generation of Parton Distributions with Uncertainties from Global QCD Analysis, *hep-ph/0201195*, 1-43.
2. Baldit, A., *et al.* (1994) Study of the isospin symmetry breaking in the light quark sea of the nucleon from the Drell-Yan process, *Phys. Lett.* **B332**, 244-250.
3. Hawker, E. A., *et al.* (1998) Measurement of the Light Antiquark Flavor Asymmetry in the Nucleon Sea, *Phys. Rev. Lett.*, **80**, 3715 -3718.
4. Ackerstaff, K., *et al.* (1998) Flavor Asymmetry of the Light Quark Sea from Semi-inclusive Deep-Inelastic Scattering, *Phys. Rev. Lett.*, **81**, 5519-5522.
5. Dressler, B., *et al.* (2000) Flavor asymmetry of polarized antiquark distributions and semi-inclusive DIS, *Eur. Phys. J.*, **C14**, 147-157.
6. Bourrely, C., *et al.* (2002) A statistical approach for polarized parton distributions, *Eur. Phys. J.* **C23**, 487-501.
7. Thomas, A. W. (1983) A limit on the pionic component of the nucleon through SU(3) flavor breaking in the sea, *Phys. Lett.* **B126**, 97-100.
8. Ackerstaff, K., *et al.* (1998) The HERMES Spectrometer, *Nucl. Inst. Meth.* **A417**, 230-265.
9. Akapov, A., *et al.* (2002) The HERMES dual-radiator ring imaging Cherenkov detector, *Nucl. Inst. Meth.* **A479**, 511-530.
10. Filippone, B. W. and Ji, X. (2001), The spin structure of the nucleon, *hep-ph/010224*, 1-70.
11. Adams, D., *et al.* (1997), Spin structure of the proton from polarized inclusive deep-inelastic muon-proton scattering, *Phys. Rev.* **D56**, 5330-5358.
12. Ackerstaff, K., *et al.* (1999) Flavor decomposition of the polarized quark distributions in the nucleon from inclusive and semi-inclusive deep-inelastic scattering, *Phys. Lett.* **B464**, 123-134.
13. Adeva, B., *et al.* (1998) Polarised quark distributions in the nucleon from semi-inclusive asymmetries, *Phys. Lett.* **B420**, 180-190.
14. Adeva, B., *et al.* (1994) Combined analysis of world data on nucleon spin structure functions, *Phys. Lett.* **B320**, 400-406.
15. Close, F. E. and Roberts. R. G. (1993) Consistent analysis of the spin content of the nucleon, *Phys. Lett.* **B316**, 165-171.

GENERALIZED PARTON DISTRIBUTIONS: WHAT DO THEY REALLY MEAN ?

ANDREAS FREUND

Institute of Theoretical Physics, University of Regensburg
Universitätsstr. 31, 93040 Regensburg, Germany

Abstract. In this article we review the concept of generalized parton distributions (GPDs), the current research status, as well as try to intuitively describe what type of physical information is contained within them.

1. Introduction

Scientists have striven for centuries to unravel the dynamics and the structures involved in the physical systems they have been investigating. From large scale structures in our universe over biological systems down to the smallest scales achievable in todays high energy experiments. At these smallest scales the questions one is trying to answer are "What is the substructure of nucleons, what are the dynamics of this substructure and what three dimensional picture of a nucleon is emerging ?".

In the theory of strong or color interactions (QCD) describing objects like nucleons, parton distribution functions (PDFs) encode long distance and thus non-perturbative information about nucleons which is precisely what we need in order to construct a dynamical as well as geometrical picture of these objects. Unfortunately, most high energy experiments analyzing nucleonic structure study inclusive reactions such as deep inelastic scattering (DIS) $e + p \to e + X$, which destroy the object they are studying. Though PDFs can be extracted from inclusive data, these PDFs are only single particle distributions precisely because the target is destroyed and thus vital information about the three diemsnional distribution of substructure is lost. Therefore they can only give a two dimensional picture of a nucleon. In order to gain insight into the three dimensional structure of nucleons one has to measure particle correlation functions which encode additional information on how the object as a whole reacts to an outside

21

E. Steffens and R. Shanidze (eds.), Spin Structure of the Nucleon, 21–35.

probe. These particle correlation functions can only be measured if the nucleon in the high energy reaction stays intact. This can only be achieved if there are no large color forces, responsible for a break up of a nucleon, occurring within the nucleon during the reaction. This requirement simply means that such a reaction has to be mediated by color neutral objects such as color singlets or at the least that color is saturated locally in the reaction. The experimental signature of such a reaction is a so called rapidity gap meaning that the produced particles or particle which are/is well localized in the detector, is clearly separated from the intact final state nucleon in the detector with no detector activity in between the two. There are many reactions of this kind such as hard diffraction $e + p \rightarrow e + p + X$ or, in particular, deeply virtual Compton scattering (DVCS) $e + p \rightarrow e + p + \gamma$ [1, 2, 3] which is the most exclusive example of hard diffraction. Hard is meant here in the sense of the presence of a large scale in the reaction such as a large momentum transfer. In the QCD description of fully exclusive hard reactions such as DVCS, we finally encounter the objects we have been looking for, namely, particle correlation functions. They appear in the factorization theorems of these reactions [4, 5]. Factorization theorems are the statement that, within QCD, one can factorize the cross section or scattering amplitude of a particular hard reaction to all orders in perturbation theory into a convolution of a finite, or infra-red safe ,hard scattering function particular to each reaction but computable to all orders in perturbation theory and a PDF which is, however, a universal object and can be used in other hard reactions, modulo contributions which are suppressed in the large scale of the reaction. These PDFs cannot be computed within perturbative QCD save for their energy dependence which is known as evolution. However, they are given as Fourier transforms of matrix elements of non-local operators. The key thing here are the in and out states of these matrix elements. In inclusive reactions such as DIS, the in and out state is the same since the scattering amplitude can be directly related through the optical theorem to a scattering reaction which has the same in and out state. In hard, exclusive reactions such as DVCS, the in and out state differ in their momenta due to a finite momentum transfer in the t-channel of the reaction onto the outgoing nucleon and therefore these PDFs depend on more variables, namely those characterizing the momentum difference, than the PDFs in inclusive reactions. The behavior of these PDFs called generalized parton distributions (GPDs) [1, 3, 6] under a change of its variables encodes the response of the entire nucleon, i.e. its substructure, to the outside probe with the nucleon staying intact. Therefore, these GPDs are particle correlation functions and a complete mapping of the GPDs in all its variables through experiments would give us, for the first time, a complete three dimensional picture of a nucleon.

In the following we will describe what is generally known about GPDs mathematically without talking about specific models and what picture of a nucleon they seem to convey. Readers not interested in the mathematical description can skip ahead to the last section.

2. What we know about GPDs

GPDs are defined by Fourier transforms of twist-two operators sandwiched between unequal momentum nucleon states (p, p' are the initial and final state nucleon momenta).The essential feature of such two parton correlation functions is the presence of a finite momentum transfer, $\Delta = p - p'$, in the t-channel. Hence the partonic structure of the hadron is tested at *distinct* momentum fractions.

Matrix elements for quark and gluon correlators of unequal momentum nucleon states may be defined in a number of ways. We choose a definition which treats the initial and final state nucleon momentum symmetrically by involving parton light-cone fractions with respect to the momentum transfer, $\Delta = p - p'$, and the average momentum, $\bar{P} = (p + p')/2$. The flavor singlet and non-singlet (S,NS) quark, and the gluon (G) matrix elements of the non-local operators, involving a light-cone vector z^μ ($z^2 = 0$), are defined by

$$
\begin{aligned}
2M_a^{NS}(\bar{P} \cdot z, \Delta \cdot z, t) &= \langle N(P_-) | \bar{\psi}_a \left(-\frac{z}{2}\right) \mathcal{P}\hat{z}\psi_a \left(\frac{z}{2}\right) | N(P_+) \rangle + \\
&\quad \langle N(P_-) | \bar{\psi}_a \left(\frac{z}{2}\right) \mathcal{P}\hat{z}\psi_a \left(-\frac{z}{2}\right) | N(P_+) \rangle, \\
2M^S(\bar{P} \cdot z, \Delta \cdot z, t) &= \sum_a \langle N(P_-) | \bar{\psi}_a \left(-\frac{z}{2}\right) \mathcal{P}\hat{z}\psi_a \left(\frac{z}{2}\right) | N(P_+) \rangle - \\
&\quad \langle N(P_-) | \bar{\psi}_a \left(\frac{z}{2}\right) \mathcal{P}\hat{z}\psi_a \left(-\frac{z}{2}\right) | N(P_+) \rangle, \\
2M^G(\bar{P} \cdot z, \Delta \cdot z, t) &= z^\mu z^\nu \langle N(P_-) | G_{\mu\rho} \left(-\frac{z}{2}\right) \mathcal{P}G_\nu^\rho \left(\frac{z}{2}\right) | N(P_+) \rangle + \\
&\quad z^\mu z^\nu \langle N(P_-) | G_{\mu\rho} \left(\frac{z}{2}\right) \mathcal{P}G_\nu^\rho \left(-\frac{z}{2}\right) | N(P_+) \rangle,
\end{aligned}
$$

$$(1)$$

where $P_{+,-} = \bar{P} \pm \frac{\Delta}{2}$, $t = \Delta^2$ is the four-momentum transfer, $\hat{z} = \gamma^\mu z_\mu$, $\bar{P} \cdot z$ and $\Delta \cdot z$ are dimensionless Lorentz scalars, $a = u, d, s, c...$ is a flavor index and the symbol $\mathcal{P}$ represents the usual path ordered exponential. The spin and isospin dependence have been suppressed for convenience. Polarized matrix elements are defined in a similar fashion, but with $\hat{z} \to \hat{z}\gamma^5$ for the quark and with the implicit metric $g^{\mu\nu} \to i\epsilon^{\mu\nu-+}$ for the gluon. The unpolarized matrix elements, and thus their associated GPDs, have definite symmetry properties. G-parity gives (suppressing the t-dependence)

$$M^{NS}(\bar{P} \cdot z, \Delta \cdot z) = -M^{NS}(-\bar{P} \cdot z, -\Delta \cdot z)$$
$$M^{S,G}(\bar{P} \cdot z, \Delta \cdot z) = M^{S,G}(-\bar{P} \cdot z, -\Delta \cdot z). \quad (2)$$

The opposite properties are observed for the polarized case. Hermitian conjugation gives for all polarized and unpolarized species ($i = S, NS, G$)

$$M^i(\bar{P} \cdot z, \Delta \cdot z) = M^i(\bar{P} \cdot z, -\Delta \cdot z). \quad (3)$$

The matrix elements can be most generally represented by a double spectral representation with respect to $\bar{P} \cdot z$ and $\Delta \cdot z$ [1, 3, 6] as follows:

$$
\begin{aligned}
M^{NS} &= \int_{-1}^{1} dx e^{-ix\bar{P}\cdot z} \int_{-1+|x|}^{1-|x|} dy e^{-iy\Delta \cdot z/2} \\
&\quad \left[\bar{U}'\hat{z}U F^{NS}(x,y,t) + \frac{iz^{\mu}\Delta^{\nu}\bar{U}'\sigma_{\mu\nu}U}{2m_N} K^{NS}(x,y,t) \right], \\
M^{G} &= \int_{-1}^{1} dx e^{-ix\bar{P}\cdot z} \int_{-1+|x|}^{1-|x|} dy e^{-iy\Delta \cdot z/2} \\
&\quad \left[\bar{U}'\hat{z}U \frac{\bar{P}\cdot z}{2} F^{G}(x,y,t) + \frac{iz^{\mu}\Delta^{\nu}\bar{U}'\sigma_{\mu\nu}U\Delta\cdot z}{4m_N} K^{G}(x,y,t) \right] \\
&\quad + \bar{U}'U\Delta \cdot z \int_{-1}^{1} dy e^{-iy\Delta \cdot z/2} D^{G}(y,t), \\
M^{S} &= \int_{-1}^{1} dx e^{-ix\bar{P}\cdot z} \int_{-1+|x|}^{1-|x|} dy e^{-iy\Delta \cdot z/2} \\
&\quad \left[\bar{U}'\hat{z}U F^{S}(x,y,t) + \frac{iz^{\mu}\Delta^{\nu}\bar{U}'\sigma_{\mu\nu}U}{2m_N} K^{S}(x,y,t) \right] \\
&\quad + \bar{U}'U\Delta \cdot z \int_{-1}^{1} dy e^{-iy\Delta \cdot z/2} D(y,t), \quad (4)
\end{aligned}
$$

where $\bar{U}'$ and U are nucleon spinors. Note that in accordance with the associated Lorentz structures, the F's correspond to helicity non-flip and the K's to helicity flip amplitudes. Henceforth, for brevity, we shall only discuss the helicity non-flip piece explicitly. However, the helicity flip case is exactly analogous. The D-terms in the last two lines permit non-zero values for the singlet M^S and M^G in the limit $\bar{P} \cdot z \to 0$ and $\Delta \cdot z \neq 0$ which is allowed by their evenness in $\bar{P} \cdot z$ (cf. eq.(2)). Conversely, M^{NS} is required by its oddness under $\bar{P} \cdot z$, for any $\Delta \cdot z$, to be zero [6]. The above spectral functions $F^i(x,y,t,Q^2)$ etc. are called double distributions and were introduced in [3] with plus momentum fractions, x, y, of the outgoing and returning partons defined as shown in the left hand plot of Fig. 1. They exist on the diamond-shaped domain shown to the right of Fig. 1.

By making a particular choice of the light-cone vector, z^μ, as a light-ray vector (so that in light-cone variables, $z_\pm = z_0 \pm z_3$, only its minus component is non-zero $z^\mu = (0, z_-, 0)$) one may reduce the double spectral representation of eq.(4), defined on the entire light-cone, to a one dimensional spectral representation, defined along a light ray, depending on the skewedness parameter, ξ, defined by

$$\xi = \Delta \cdot z/2\bar{P} \cdot z = \Delta_+/2\bar{P}_+ . \tag{5}$$

Thus the outgoing parton lines only have a single plus momentum relative to any particular external momenta, therefore the GPDs are related to these DDs via a reduction integral, involving $\delta(v - x - \xi y)$, along the off-vertical lines in the diamond (the dotted line corresponds to $v = \xi$):

$$H^i(v, \xi) = \int_{-1}^{1} dx \int_{-1+|x|}^{1-|x|} dy \, \delta(x + \xi y - v) F^i(x, y). \tag{6}$$

GPDs were originally introduced in [1]. Here we use, for convenience, Ji's *off-forward* parton distributions for quarks and gluons [2] that can be probed in hard exclusive processes, for example the unpolarized helicity-preserving distributions, $H^{q,g}(v, \xi)$, with dependent variable $v \in [-1, 1]$ and skewedness $\xi = x_{bj}/(2 - x_{bj})$. From the Lorentz structure of their definitions one can derive polynomiality conditions on their moments:

$$M_N = \int_{-1}^{1} dv v^{N-1} [H^q(v, \xi) + H^g(v, \xi)] = \sum_{k=0}^{N/2} \xi^{2k} C_{2k,N} . \tag{7}$$

The quark distribution may be expressed as a sum of odd and even functions (about the point $v = 0$) which correspond to the singlet and non-singlet distributions, respectively. In Ji's definition the gluon distribution is purely odd about $v = 0$. This implies that odd moments are integrals over the non-singlet quark distribution and the even moments are integrals over the sum of the quark singlet and gluon distributions. For example, the first four moments are given by

$$M_1 = \int_0^1 dv H^{q,NS}(v, \xi) = C_{0,1} , \tag{8}$$

$$M_2 = \int_0^1 dv v [H^{q,S}(v, \xi) + H^g(v, \xi)] = C_{0,2} + \xi^2 C_{2,2} , \tag{9}$$

$$M_3 = \int_0^1 dv v^2 H^{q,NS}(v, \xi) = C_{0,3} + \xi^2 C_{2,3} , \tag{10}$$

$$M_4 = \int_0^1 dv v^3 [H^{q,S}(v,\xi) + H^g(v,\xi)] = C_{0,4} + \xi^2 C_{2,4} + \xi^4 C_{4,4} \quad (11)$$

where M_1 and M_2 are generalizations of the number density and momentum sum rules (which implies that $C_{0,1} = 3$, for three valence quarks, and $C_{0,2} = 1$ from momentum conservation).

For the numerical solution of the renormalization group equations (RGEs in [7] the *natural* off-diagonal PDFs, $\mathcal{F}^i(X,\zeta)$, defined by Golec-Biernat and Martin [8] were preferred. They depend on the momentum fraction $X \in [0,1]$ of the incoming proton's momentum, p, and the skewedness variable $\zeta = \Delta^+/p^+ = 2\xi/(1+\xi)$ (so that $\zeta = x_{bj}$ for DVCS). For the quark case, the relationship of the quark and anti-quark distributions, $\mathcal{F}^q(X,\zeta), \mathcal{F}^{\bar{q}}(X,\zeta)$, to Ji's single function $H^q(v,\xi)$ is shown in Fig. 2. More explicitly, for $v \in [-\xi,1]$:

$$\mathcal{F}^{q,a}\left(X = \frac{v+\xi}{1+\xi},\zeta\right) = \frac{H^{q,a}(v,\xi)}{1-\zeta/2}, \quad (12)$$

and for $v \in [-1,\xi]$

$$\mathcal{F}^{\bar{q},a}\left(X = \frac{\xi-v}{1+\xi},\zeta\right) = -\frac{H^{q,a}(v,\xi)}{1-\zeta/2}. \quad (13)$$

The two distinct transformations between v and X for the quark and anti-quark cases are shown explicitly on the left hand side of Eqs.(12, 13). There are two distinct regions: the DGLAP region, $X > \zeta$ ($|v| > \xi$), in which the GPDs obey a generalized form of the DGLAP equations for PDFs, and the ERBL region, $X < \zeta$ ($|v| < \xi$), where the GPDs obey a generalized form of the ERBL equations for distributional amplitudes. In the ERBL region, due to the fermion symmetry, $\mathcal{F}^q$ and $\mathcal{F}^{\bar{q}}$ are not independent. In fact $\mathcal{F}^q(X,\zeta) = -\mathcal{F}^{\bar{q}}(\zeta - X,\zeta)$, which leads to an anti-symmetry of the unpolarized quark singlet distributions (summed over flavor a), $\mathcal{F}^S = \sum_a \mathcal{F}^{q,a} + \mathcal{F}^{\bar{q},a}$, about the point $\zeta/2$ (the non-singlet and the gluon, $\mathcal{F}^g$, which is built from $v H^g_{Ji}(v,\xi)$, are symmetric about this point).

By design, the moments of GPDs based on double distributions are automatically polynomials in the skewedness variable, ξ. This can be seen by taking moments of both sides of Eq. (6) and by using the delta function to perform the integration over v on the right hand side:

$$\int_{-1}^1 dv v^{N-1} H(v,\xi) = \int_{-1}^1 dx \int_{-1+|x|}^{1-|x|} dy \, (x+\xi y)^{N-1} F_{DD}(x,y) \equiv$$

$$\sum_{k=0}^{N-1} \xi^k \binom{N-1}{k} \int_{-1}^1 dx \int_{-1+|x|}^{1-|x|} dy \, F_{DD}(x,y) \, x^{N-1-k} y^k. \quad (14)$$

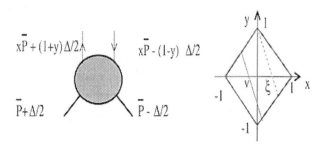

Figure 1. Symmetric double distributions (left), indicating momentum fractions of the outgoing and returning partons, and (right) their physical domain.

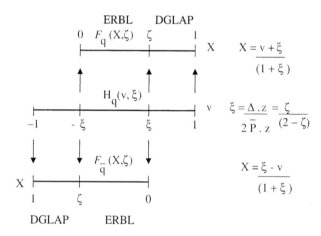

Figure 2. The relationship between $\mathcal{F}^q(X,\zeta)$, $\mathcal{F}^{\bar{q}}(X,\zeta)$ and Ji's function $H^q(v,\xi)$ with $v \in [-1, 1]$ and $X \in [0, 1]$.

Performing the above integrals for a particular model for $F_{DD}(x,y)$ determines the respective coefficients in the polynomial. In order to ensure non-zero coefficients, $C_{N,N}$ of the highest power of ξ^N, for even N, one has to include an additional term, the so-called 'D-term' [6], to both the quark singlet and gluon GPD (cf. Eq. (7)). The missing (odd) powers of ξ reflect the symmetry of the matrix elements under $\xi \to -\xi$ [11].

The D-term can be computed in the chiral-quark soliton model [6, 12] to be given as a truncated expansion in terms of odd Gegenbauer polynomials:

$$H^D(v,\xi) = \Theta(\xi - |v|) D\left(a = \frac{v}{\xi}, t = 0\right) / N_F,$$

$$D(a) = \left(1 - a^2\right)\left[-4.0 C_1^{3/2}(a) - 1.2 C_3^{3/2}(a) - 0.4 C_5^{3/2}(a)\right], \quad (15)$$

however its influence becomes negligible in the small ξ (or ζ) limit (see e.g. Fig. 7 of [7]). The restriction of the D-term to the ERBL region and the fact that it is restricted to odd functions of v/ξ guarantees that upon integration it provides the missing highest coefficient, $C_{N,N}$, for even N, without modifying the coefficients of other powers of ξ.

In the following we will discuss the behavior of GPDs under a change in scale μ^2 which is governed by a RGE or evolution equation directly derivable from a factorization theorem or a non-local OPE [1]. There are two sets of RG equations, one for the DGLAP and one for the ERBL region, since the GPDs behave qualitatively different in these two regions. Note that the evolution in the ERBL region depends on the DGLAP region (see eq. (19)) whereas the DGLAP evolution is independent of the ERBL region. This is natural since evolution always leads to a degradation of the initial parton momentum towards lower values due to the parton splitting enoced within the evolution kernels, however never the other way around. Therefore, all partons eventually accumulate in the ERBL region. Furthermore the evolved functions remain continuous under evolution and continue to satisfy all appropriate symmetries in the ERBL region. In the following discussion we only use the $\mathcal{F}$'s for convenience.

In the DGLAP region the singlet and gluon distributions mix under evolution:

$$
\begin{aligned}
\frac{d\mathcal{F}^S(y,\zeta,Q^2)}{d\ln(Q^2)} &= \int_y^1 \frac{dz}{z} P_{qq}\left(\frac{y}{z},\frac{\zeta}{z}\right)_+ \mathcal{F}^S(z,\zeta,Q^2) \\
&+ \left(1-\frac{\zeta}{2}\right)\int_y^1 \frac{dz}{z} P_{qg}\left(\frac{y}{z},\frac{\zeta}{z}\right) \mathcal{F}^G(z,\zeta,Q^2)\,, \\
\frac{d\mathcal{F}^G(y,\zeta,Q^2)}{d\ln(Q^2)} &= \int_y^1 \frac{dz}{z} P_{gg}\left(\frac{y}{z},\frac{\zeta}{z}\right)_+ \mathcal{F}^G(z,\zeta,Q^2) \\
&+ \frac{1}{1-\frac{\zeta}{2}}\int_y^1 \frac{dz}{z} P_{gq}\left(\frac{y}{z},\frac{\zeta}{z}\right) \mathcal{F}^S(z,\zeta,Q^2)\,, \quad (16)
\end{aligned}
$$

with generalized DGLAP kernels [9]. The NS combinations do not mix under evolution so we omit them for brevity.

The +-distribution for the DGLAP kernel, regulates the divergence as $z \to y$. In general it is defined as follows:

$$
P(x,\zeta)_+ = P(x,\zeta) - \delta(x-1)\int_0^1 dx' P(x',\zeta). \quad (17)
$$

The numerical implementation of the +-distribution applied to the integrals in eq.(16) is as follows:

$$
\int_y^1 \frac{dz}{z} P\left(\frac{y}{z},\frac{\zeta}{z}\right)_+ \mathcal{F}(z,\zeta) = \int_y^1 \frac{dz}{z} P\left(\frac{y}{z},\frac{\zeta}{z}\right)(\mathcal{F}(z,\zeta)-\mathcal{F}(y,\zeta))
$$

$$-\mathcal{F}(y,\zeta)\Big[\int_{\frac{y}{\zeta}}^{1} dz P\left(z,\frac{\zeta}{y}\right) - \int_{y}^{1}\frac{dz}{z}P\left(z,z\frac{\zeta}{y}\right)$$

$$+\frac{y}{\zeta}\int_{0}^{\zeta/y} dz V\left(z\frac{y}{\zeta},\frac{y}{\zeta}\right)\Big]. \tag{18}$$

Note that a generalized ERBL kernel appears in the + regulrization of the DGLAP kernels!

In the ERBL region we have

$$\frac{d\mathcal{F}^{S}(y,\zeta,Q^{2})}{d\ln(Q^{2})} = \Big[\int_{y}^{1}\frac{dz}{\zeta}V^{qq}\left(\frac{y}{\zeta},\frac{z}{\zeta}\right)_{+} + \int_{0}^{y}\frac{dz}{\zeta}V^{qq}\left(\frac{\overline{y}}{\zeta},\frac{\overline{z}}{\zeta}\right)_{+}$$

$$\mp \int_{\zeta}^{1}\frac{dz}{\zeta}V^{qq}\left(\frac{\overline{y}}{\zeta},\frac{z}{\zeta}\right)\Big]\mathcal{F}^{S} + \left(1-\frac{\zeta}{2}\right)\Big[\int_{y}^{1}\frac{dz}{\zeta^{2}}V^{qg}\left(\frac{y}{\zeta},\frac{z}{\zeta}\right)$$

$$-\int_{0}^{y}\frac{dz}{\zeta^{2}}V^{qg}\left(\frac{\overline{y}}{\zeta},\frac{\overline{z}}{\zeta}\right) \mp \int_{\zeta}^{1}\frac{dz}{\zeta^{2}}V^{qg}\left(\frac{\overline{y}}{\zeta},\frac{z}{\zeta}\right)\Big]\mathcal{F}^{G},$$

$$\frac{d\mathcal{F}^{G}(y,\zeta,Q^{2})}{d\ln(Q^{2})} = \Big[\int_{y}^{1}\frac{dz}{\zeta}V^{gg}\left(\frac{y}{\zeta},\frac{z}{\zeta}\right)_{+} + \int_{0}^{y}\frac{dz}{\zeta}V^{gg}\left(\frac{\overline{y}}{\zeta},\frac{\overline{z}}{\zeta}\right)_{+}$$

$$\pm \int_{\zeta}^{1}\frac{dz}{\zeta}V^{gg}\left(\frac{\overline{y}}{\zeta},\frac{z}{\zeta}\right)\Big]\mathcal{F}^{G} + \frac{1}{\left(1-\frac{\zeta}{2}\right)}\Big[\int_{y}^{1}dz V^{gq}\left(\frac{y}{\zeta},\frac{z}{\zeta}\right)$$

$$-\int_{0}^{y}dz V^{gq}\left(\frac{\overline{y}}{\zeta},\frac{\overline{z}}{\zeta}\right) \pm \int_{\zeta}^{1}dz V^{gq}\left(\frac{\overline{y}}{\zeta},\frac{z}{\zeta}\right)\Big]\mathcal{F}^{S}, \tag{19}$$

with generalized ERBL kernels [9], where the bar notation means for example, $\overline{z/\zeta} = 1 - z/\zeta$. The upper signs correspond to the unpolarized case and the lower signs to the polarized case.

The numerical implementation of the +-distribution, again applied to the whole kernel, takes the following form in the ERBL region:

$$\int_{y}^{1}\frac{dz}{\zeta}V\left(\frac{y}{\zeta},\frac{z}{\zeta}\right)_{+}\mathcal{F}(z,\zeta) = \int_{y}^{1}\frac{dz}{\zeta}V\left(\frac{y}{\zeta},\frac{z}{\zeta}\right)[\mathcal{F}(z,\zeta) - \mathcal{F}(y,\zeta)]$$

$$+\mathcal{F}(y,\zeta)\Big[\int_{y}^{\zeta}\frac{dz}{\zeta}\left(V\left(\frac{y}{\zeta},\frac{z}{\zeta}\right) - V\left(\frac{\overline{z}}{\zeta},\frac{\overline{y}}{\zeta}\right)\right)$$

$$+\int_{\zeta}^{1}\frac{dz}{\zeta}V\left(\frac{y}{\zeta},\frac{z}{\zeta}\right)\Big],$$

$$\int_{0}^{y}\frac{dz}{\zeta}V\left(\frac{\overline{y}}{\zeta},\frac{\overline{z}}{\zeta}\right)_{+}\mathcal{F}(z,\zeta) = \int_{0}^{y}\frac{dz}{\zeta}V\left(\frac{\overline{y}}{\zeta},\frac{\overline{z}}{\zeta}\right)[\mathcal{F}(z,\zeta) - \mathcal{F}(y,\zeta)] +$$

$$\mathcal{F}(y,\zeta)\int_{0}^{y}\frac{dz}{\zeta}\Big[V\left(\frac{\overline{y}}{\zeta},\frac{\overline{z}}{\zeta}\right) - V\left(\frac{z}{\zeta},\frac{y}{\zeta}\right)\Big] \tag{20}$$

where the terms have been arranged in such away that all non-integrable divergences explicitly cancel in each term separately[10].

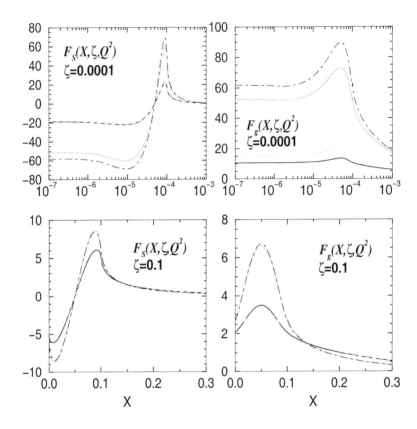

Figure 3. Unpolarized NLO input and evolved singlet quark and gluon GPDs at small and large skewedness. The solid curves are the input MRSA' GPDs at $Q_0 = 2$ GeV, the dotted ones show them evolved to $Q = 10$ GeV. The dashed curves are input GRV98 GPDs at $Q_0 = 2$ GeV and the dashed-dotted ones show them evolved to $Q = 10$ GeV. The quark singlet is scaled by a factor of 10^{-4} at $\zeta = 0.0001$ and by 10^{-2} at $\zeta = 0.1$. For $\zeta = 0.1$ the symmetry of the gluon GPD and the anti-symmetry of the singlet quark GPD are apparent about the point $X = \zeta/2 = 0.05$.

Below we demonstrate the effect of evolution on two sets of input distributions, here for the unpolarized case only. As can be seen from the two figures: first the symmetries are kept under evolution as it should be and the relative change in going from LO to NLO is moderate.

We will not discuss models of double distributions and GPDs and how they compare to experimental data on DVCS. A good summary of the current status on modeling and comparison of GPDs with data can be found in [13].

Now that we have explored all the mathematical features of GPDs it is high time to ask what the physical picture is that they convey ?

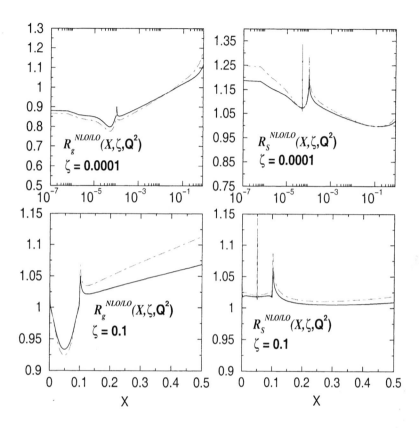

Figure 4.   The ratio of NLO to LO evolved GPDs for the unpolarized (upper plot, MRSA') The solid and dashed-dotted curves shows the ratio at $Q = 5$ GeV and $Q = 10$ GeV , respectively.

What is the physical picture behind GPDs ?

DVCS (see Fig. 5) is the prime process in which to measure GPDs. The reason for this is quite simple. With a real photon one has an elementary particle in the final state rather than a bound state like a vector meson or an even more complicated state like several mesons or hadrons. Therefore, the factorization theorem for the DVCS scattering amplitude [3, 4] is merely a simple convolution of a hard scattering function with only *one* GPD rather than with other unknown non-perturbative functions as in the case of vector meson production.

The t integrated DVCS cross section itself is given by

$$\sigma_{\text{DVCS}}(\gamma^* p \to \gamma p) = \frac{\alpha^2 x_{bj}^2 \pi}{Q^4 B} |\mathcal{T}_{\text{DVCS}}|^2 |_{t=0}, \tag{21}$$

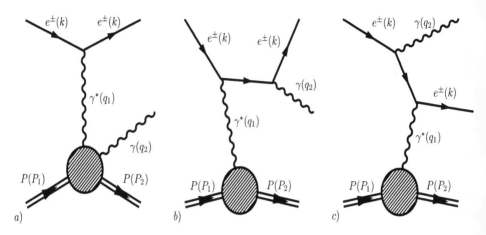

Figure 5. a) DVCS graph, b) Bethe-Heitler with photon from final state lepton and c) with photon from initial state lepton.

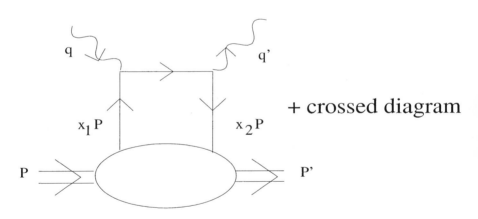

Figure 6. LO handbag diagram for DVCS. Here $x_1 = X$ and $x_2 = X - \zeta$.

with B the slope of the t-dependence and the DVCS amplitude $\mathcal{T} \simeq Im\mathcal{T} \propto \mathcal{F}(\zeta, \zeta, Q^2)$ in LO. This is true up to a $\zeta = x_{bj} \simeq 0.2 - 0.3$. Hence DVCS is dominated, at least in a very broad region of phase space, by the crossover point between the DGLAP and ERBL region. At this particular point in phase space, $X = \zeta$, the parton line carrying momentum fraction x_2 in Fig. 6 is becoming "soft" and all the momentum is carried by the incoming quark with fraction x_1. Factorization for DVCS [4] still holds in this situation, however, the point $X = \zeta$ in the GPD is rather peculiar. One should recall that the GPD is defined by a Fourier transform of a non-local matrix element on a light ray (in terms of the DD's this means the introduction of a δ-function in the reduction integral eq. (6)) and that the Fourier conjugate variables are the light-ray distance z^- between operators on the light ray and a momentum variable, here either X or $X - \zeta$. This means that for $X - \zeta \to 0$, $z^- \to \infty$ and therefore the operators have an infinite separation on the light ray or more physically speaking that there is no resolution of the probed object in the $-$ direction. This situation is analogous to inclusive DIS in the limit of $x_{bj} \to 0$. Thus inclusive scattering at small x_{bj} and DVCS up to a large x_{bj} in the valance region is dominated by the same type of particle configurations with the only difference being that the configurations in DVCS remain correlated since the proton stays intact! What does this all mean ?

To put it simply: the particle configurations which dominate the DVCS cross section are much bigger, in their extension on the light ray, than the probed object itself! Therefore, even deep into the valance region, one is not probing the actual quark structure of the proton but rather the QCD vacuum as influenced by and interacting with the three valance quarks. Moreover, in varying t which has its main component in the transverse plane perpendicular to the light ray, one starts to map the three dimensional structure of the QCD vacuum or the virtual pion cloud of the proton as one would have said in earlier days.

There is a very intuitive picture of why this interpretation is indeed true and one is not really probing the valance structure of the proton within DVCS. The main argument is simply that the proton stays intact! Consider the following situation: The proton is moving, in momentum space, along the + direction of the lightcone i.e. has its momentum in the positive 3- or $+z$-direction with each quark carrying about a third of the total momentum, $X \simeq 1/3$. If a valance quark were to be struck by a virtual photon which has large $+ \, (-x_{bj}P_+ \simeq -P_+/3)$ and $- \, (Q^2/2x_{bj}P_+ \simeq 3Q^2/2P_+)$ components with $P_+ \simeq O(Q)$, the valance quark would then only have a large $-$ component but a quasi zero $+$ component since $X \simeq x_{bj}$. This means that the struck valance quark would have a large momentum in the $-z$-direction, opposite to that of the other two quarks, radiate a real photon which is then moving in the $-z$-direction, creating the large rapidity gap (the collinear quarks and proton move in opposite directions) necessary for this type of reaction, and then becoming "soft" i.e. with no large momentum components. The transition matrix element i.e. the overlap integral, between an initial Fock state with three collinear or "fast" quarks to a final Fock state with two collinear or "fast" quarks and one soft or "slow" quark is exponentially suppressed since the probability of two collinear and one "soft" quark forming a proton in the final state is exponentially suppressed with the relative lightcone distance of the collinear quarks and the "soft" quark. However, DVCS is observed at large x_{bj} [14], therefore, the only alternative picture is the one where the virtual photon is not scattering on a valance quark but rather on a q or $\bar{q}$ from the sea i.e. the QCD vacuum where either the q or the $\bar{q}$ has a large + momentum fraction matching the one from the virtual photon. In other words, either the struck q or the struck $\bar{q}$ starts to move in the $-z$ direction and then annihilates with a "soft" $(X - \zeta \simeq 0)$ $\bar{q}$ or q from the sea into a real photon with large $-$ momentum moving along the $-z$ direction as it should. None of the valance quarks is involved in the reaction and therefore it is not very difficult for the proton to stay intact. Note, however, that the sea configurations probed at large x_{bj} in DVCS cannot directly be probed in inclusive DIS, since it involves too large a lightcone distance as compared to the one allowed in an inclusive reaction at large x_{bj} (large x_{bj} in inclusive reactions means a smaller allowable lightcone distance of the operators). Thus there exist additional QCD vacuum configurations which can only be probed in hard exclusive reactions like DVCS.

To summarize once more, the region of a GPD around $X \simeq \zeta$ or $X \simeq 0$ which is related to $X \simeq \zeta$ via symmetry, corresponds to large lightcone distances and therefore has nothing to do with the direct valance structure of the proton but rather with the structure of the sea i.e. the QCD vacuum. However, if one were to consider other reactions like vector meson production, the situation changes since one does not want to produce an elementary particle which has no size and therefore allows particle configuration of "infinite" extent in its creation but rather a bound state with a finite size. In keeping with the above picture, the configurations which would be most dominant in vector meson production are the ones which resolve the produced meson. Therefore, the smaller/larger the object the more/less one is dominated by configurations or correlations in the GPD near $X \simeq \zeta$ i.e. an Υ probes a GPD more like a real photon than a ρ does. Or put differently, the "bigger" the produced object the further away from the point $X = \zeta$ one probes the GPD i.e. at large x_{bj} more of the valance structure is revealed in pion production than in J/ψ production. Experiments bear out this picture: At large Q^2, ρ's are more easily produced at large x_{bj} than J/ψ's, whereas both are more or less equally produced at small x_{bj} All of this has interesting ramifications for experiments with nuclei, small x_{bj} etc. which, however, goes beyond the scope of this article.

In closing, on can truly say that GPDs do indeed give us a new, very intriguing and intuitive three dimensional physical picture of the structure of hadrons. However, it depends on the type of produced final state that selects which structures within the hadron are revealed and which are not. Thus one needs to understand the question one asks first before on can understand the answer they provide us.

References

1. D. Müller *et al.*, Fortschr. Phys. **42**, 101 (1994).

2. X. Ji, Phys. Rev. D **55** 7114 (1997); J. Phys. G **24**, 1181 (1998).
3. A. V. Radyushkin, Phys. Rev. D **56**, 5524 (1997).
4. J. C. Collins and A. Freund, Phys. Rev. D **59**, 074009 (1999).
5. J. C. Collins, L. Frankfurt, M. Strikman, Phys. Rev. D 56 (1997) 2982.
6. M. V. Polyakov and C. Weiss, Phys. Rev. D **60**, 114017 (1999).
7. A. Freund and M. McDermott, Phys. Rev. D **65**, 074008 (2002).
8. K. J. Golec-Biernat and A. D. Martin, Phys. Rev. D **59**, 014029 (1999).
9. A. V. Belitsky, A. Freund and D. Müller, Nucl. Phys. B574 (2000) 347.
10. The endpoint ($z = 1$) concentrated terms in eqs.(16,19) can be found in [9] and are determined by the requirement that the zeroth order conformal moment of the kernel has to be zero. Note that in the numerical implementation of the evolution the pure singlet component in the qq kernel in both DGLAP and ERBL regions is not regularized (for more details see [9]).
11. The fact that $\pi(x, y) = \pi(x, -y)$ guarantees that odd powers of ξ are missing from the moments of $H(v, \xi)$. This is required by hermiticity and time reversal of the corresponding hadronic matrix elements.
12. N. Kivel, M. V. Polyakov and M. Vanderhaeghen, Phys. Rev. D **63**, 114014 (2001).
13. A. Freund, M. McDermott amd M. Strikman, hep-ph/0208160.
14. HERMES Collaboration, A. Airapetian *et al.*, Phys. Rev. Lett. **87**, 182001 (2001).

PHYSICS OBJECTIVES FOR FUTURE STUDIES
OF THE SPIN STRUCTURE OF THE NUCLEON

WOLF-DIETER NOWAK

DESY Zeuthen, Platanenallee 6, D-15738 Zeuthen, Germany
e-mail: Wolf-Dieter.Nowak@desy.de

Abstract. Physics perspectives are shown for future experiments in electron or positron scattering on nucleons, towards a deep and comprehensive understanding of the angular momentum structure of the nucleon in the context of Quantum Chromodynamics. Measurements of Generalised Parton Distributions in exclusive reactions and precise determinations of forward Parton Distributions in semi-inclusive deep inelastic scattering are identified as major physics topics. Requirements are discussed for a next generation of high-luminosity fixed-target experiments in the energy range 30-200 GeV.

1. Introduction

Charged leptons have been used for more than two decades as a very powerful tool for studying the *momentum structure* of the nucleon, in a wide variety of experimental approaches. More recently, high-energy polarised beams and high-density polarised targets have become accessible and have proven to be indispensable tools in studying the *angular momentum structure* of the nucleon.

At large enough Q^2, the four-momentum-transfer squared of the photon mediating the lepton-nucleon interaction, short-range phenomena ('hard' photon-parton interactions) are successfully described by perturbative Quantum Chromodynamics (QCD). In contrast, long-range ('soft') phenomena and especially parton correlations in hadrons, over a broad range in Q^2, are still lacking a satisfactory theoretical description. Here it may be expected that contemporary theoretical developments will at some point turn into a calculable field theoretical description of hadronic structures.

E. Steffens and R. Shanidze (eds.), Spin Structure of the Nucleon, 37–49.

The spin of the nucleon as a whole is $\frac{\hbar}{2}$, irrespectively of the resolving power Q^2 of the virtual photon. Due to angular momentum conservation in QCD the individual contributions of the parton's spins and orbital angular momenta always add up to this value, although they vary considerably with Q^2. Precise measurements at high and especially moderate Q^2 are required to be able to reliably 'bridge' into the 'critical' region of $Q^2 \leq 0.5$ GeV^2 to eventually test theoretical descriptions of soft phenomena.

In this paper, based on recent theoretical developments in the field, an experimentalist's perspective is given on physics prospects for possible future electron-nucleon fixed-target experiments in the center-of-mass energy range of up to a few tens of GeV^2, with high resolution, high luminosity and polarised beams and/or targets.

2. Overview of the Relevant Quantities

The cross section of the inclusive reaction $eN \rightarrow eX$ is not yet exactly calculable from theory. Instead it is presently parameterised by nonperturbative 'structure functions' which in turn, in the framework of the quark-gluon picture of the nucleon, are expressed in terms of Parton Distribution Functions (PDFs). These functions have proven to be very useful in the description of the momentum and spin structure of partons. However, they do not yet include the parton's *orbital* angular momenta which are considered to be essential among the various components that eventually make up the half-integer spin of the nucleon. This striking deficiency is cured in the formalism of 'Generalized Parton Distributions' (GPDs) [1, 2, 3, 4] which reached the level of practical applications only recently. This theoretical framework is capable of simultaneously treating several types of processes ranging from inclusive to hard exclusive scattering. Exclusive lepton-nucleon scattering is 'non-forward' in nature since the photon initiating the process is virtual and the final state particle is usually real, forcing a small but finite momentum transfer to the target nucleon. While it is very appealing that GPDs embody forward ('ordinary') PDFs as well as Nucleon Form Factors as limiting cases, they clearly contain a wealth of information well beyond these, notably about the hitherto unrevealed orbital angular momenta of partons (see below).

The scheme presented in Fig. 1 is intended as an easy-to-read overview of the various quantities that are considered   relevant for a comprehensive

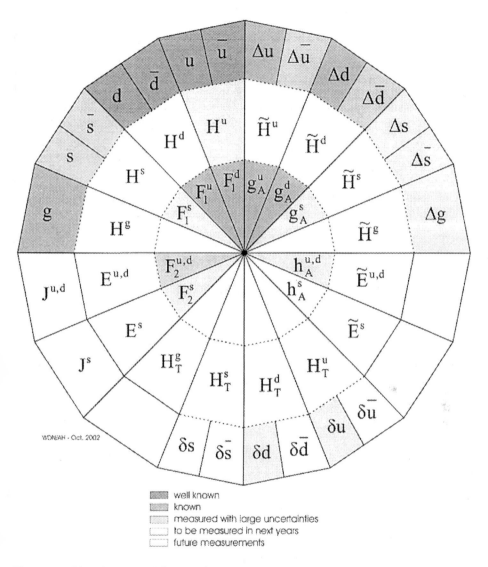

Figure 1.   Visualisation of (most of) the relevant Generalised Parton Distributions and their limiting cases, forward Parton Distributions and Nucleon Form Factors. Different shades of grey illustrate the status of their experimental access (see legend). For explanations see text.

description of the angular momentum structure of the nucleon. It is based on the formalism of GPDs which are placed in the middle of three concentric rings. The two limiting cases are located in the adjacent rings: Nucleon Form Factors, the first moments of their appropriate GPDs, are shown in the innermost ring, while PDFs, their 'forward' limits, are located in the

outermost ring. Today's experimental knowledge of the different functions
is illustrated in shades of grey, from light (no data exist) to dark (well
known).The emphasis in Fig. 1 is placed on the physics message and not on
completeness; some GPDs have been omitted. Empty sectors mean that the
function does not exist, decouples from observables in the forward limit, or
no strategy is known for its measurement. GPDs will be discussed in detail in the next chapter, PDFs in chapter 4, and Nucleon Form Factors in
chapter 5.

3. Generalized Parton Distributions

Generalized Parton Distributions depend on a resolution scale Q^2, two longitudinal momentum fractions (x, ξ) and t, the momentum transfer at the
nucleon vertex. The 'skewedness' ξ parameterises the longitudinal momentum difference of the two partons involved in the interaction (cf. Fig. 2).
Through their dependence on ξ the GPDs carry information about correlations between two different partons in the nucleon. Note that in the
'forward' limit $\xi = 0$ and $t = 0$.

For each parton species[1] f there exist four GPDs $(H^f, \tilde{H}^F, E^f, \tilde{E}^f)$ that
do not flip the *parton* helicity and additional four $(H_T^f, \tilde{H}_T^f, E_T^f, \tilde{E}_T^f)$ that
do flip it [5, 6]. In terms of chirality the two sets of *quark* GPDs can be also
referred to as chirally-even and chirally-odd ones, respectively, while gluons
do not have chirality. The *nucleon* helicity, on the other hand, is conserved
by the H-functions, but not by the E-functions. Finally, the chirally-even,
i.e. parton-helicity non-flip quark GPDs can also be classified into unpolarised (H^f, E^f) and polarised $(\tilde{H}^f, \tilde{E}^f)$ ones. The nucleon-helicity conserving quark GPDs H^f and $\tilde{H}^f$ have as forward limits the well-known
unpolarised and longitudinally polarised quark PDFs $q_f, \bar{q}_f$ and $\Delta q_f, \Delta \bar{q}_f$,
respectively. In the case of gluons, g and Δg are the forward limits of H^g
and $\tilde{H}^g$, respectively. The nucleon-helicity non-conserving GPDs E^f and
$\tilde{E}^f$ decouple from observables in the forward limit, i.e. they have no corresponding PDFs. As forward limit of the *parton-helicity flip* quark GPDs H_T^f
the transversity PDFs $\delta q_f, \delta \bar{q}_f$ are obtained, while gluons have no transversity [7].

The recent strong interest in GPDs was stimulated by the finding [2] that
the sum of the unpolarised chirally-even GPDs, $\frac{1}{2}(H^f + E^f)$, carries information about the *total* angular momentum J^f of the parton species f

---

[1]Here f stands for the quark flavors u, d, s, but here also for the gluon index g.

in the nucleon. In the limit of vanishing t the second moment of this sum approaches J^f, measured at the given Q^2. The total angular momentum carried by quarks of all flavors, $J^q = \sum_f J^f$, and that carried by gluons, J^g, are not known yet. The same holds for the corresponding orbital momenta, L^q and L^g. Through the simple integral relation $L^q = J^q - \frac{1}{2}\Delta\Sigma$, with $\frac{1}{2}\Delta\Sigma$ being the quark's *spin* contribution (cf. next chapter), a measurement of J^q would allow to determine L^q. Note that there is an ongoing unresolved controversy in the literature about the appropriate definition of angular momentum operators for quarks and gluons [8].

GPDs can in principle be revealed from measurements of various cross sections and spin asymmetries in several exclusive processes. However, there is no doubt that for a determination of individual GPDs from experimental cross sections and asymmetries a rather complicated procedure will have to be developed. Unlike in the case of forward PDFs a direct extraction of GPDs appears presently not feasible. The usual deconvolution procedure can not be applied, because the involved momentum fraction x is an entirely internal variable, i.e. it is always integrated over in the process amplitudes. A principal way to circumvent this problem by distinguishing the $log\,Q^2$ behaviour from the $1/Q^2$ behaviour [9] appears hard to realise even under substantially improved experimental conditions in the future. Hence an iteration process based on the comparison of theoretical model GPDs to experimental data seems to be the unavoidable choice. This most probably will require the combination of results from various reaction channels into a 'global fit' of the involved GPDs. Clearly, further theoretical work on the sensitivity of experimentally available quantities to different properties of the model functions, especially higher orders (see e.g. Ref. [10]) and higher twists (see e.g. Ref. [11]), seems to be of great importance.

The chirally-even GPDs ($H^f, \tilde{H}^f, E^f, \tilde{E}^f$) for u and d-quarks, which are presently the most intensely discussed functions, can be accessed experimentally in several reactions. Deeply Virtual Compton Scattering (DVCS), $ep \rightarrow ep\gamma$ (cf. left panel of Fig. 2), is presently considered the cleanest one. In this reaction the first and, so far, the only GPD-related experimental results were published recently. Note that there is still no data available that can be related to parton-helicity non-flip *gluon* GPDs. They can in principle be accessed in DVCS and meson production at small ξ, i.e. preferentially at large center-of-mass energies.

A $\sin\phi$ behaviour[2] is predicted [12] for the azimuthal dependence of the sin-

---

[2]Here ϕ is the azimuthal angle of the produced real photon around the direction of the virtual photon, relative to the lepton scattering plane.

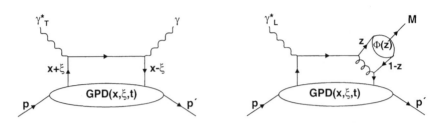

Figure 2. Illustration of the two major types of hard exclusive processes to extract GPDs: DVCS and Meson Production.

gle *beam-spin asymmetry* in the DVCS cross section. This prediction was recently proven at two different center-of-mass energies, by HERMES [13] and CLAS [14]. The asymmetry is given by the *imaginary* part of the interference term between the DVCS and Bethe-Heitler amplitudes. The involved combination of GPDs contains $H^f, \tilde{H}^f$ and E^f, where H^f is the dominant function driven by kinematical factors (cf. Ref. [15]). Access to the *real* part of the same interference term is opened by measuring the *beam-charge asymmetry* in DVCS. First results on this observable were presented by HERMES [16] very recently and confirmed the predicted $\cos\phi$ behaviour. It has to be noted that the imaginary part is probed at the special argument (ξ, ξ, t) which constitutes a second independent 'slice' in the (x, ξ, t)-plane, in addition to the 'forward slice' $(x, 0, 0)$. The x-dependence away from the line $x = \xi$ is contained in the principal value integral of the real part [17].

Information on chirally-even GPDs can also be accessed in Deeply Virtual Exclusive production of pseudoscalar and vector Mesons (DVEM), $e\,p \rightarrow e\,p\,M$ (cf. right panel of Fig. 2). While DVCS is suppressed in comparison to meson production by the additional electromagnetic coupling, the latter is suppressed by a factor $1/Q^2$ over the former. In fixed-target exclusive meson production, this results in an increase by a factor of about 10 in count rate, as compared to DVCS. On the other hand, compared to the outgoing real photon in DVCS the exclusively produced meson introduces one more complication, namely the distribution amplitude $\Phi(z)$ as an additional unknown. Here z is the momentum fraction carried by the meson. Cross sections and spin asymmetries for different channels are described by different sets of GPDs (for more details see e.g. Ref. [17]). The comparison of pseudoscalar and vector meson production reveals that the final state meson acts in fact as a helicity filter. Fig. 3 illustrates which of the chirally-even GPDs can be accessed in which reaction channels.

DVCS	$\gamma^* p \longrightarrow \gamma\, p$	H $\tilde{\text{H}}$ E $\tilde{\text{E}}$
exclusive pseudoscalar meson production	$\gamma^* p \longrightarrow \pi^0\, p$ $\gamma^* p \longrightarrow \pi^+ n$	$\tilde{\text{H}}$ $\tilde{\text{E}}$
exclusive vector meson production	$\gamma^* p \longrightarrow \rho^0\, p$ $\gamma^* p \longrightarrow \omega\, p$ $\gamma^* p \longrightarrow \phi\, p$	H E

Figure 3. *Examples of hard exclusive processes and the involved chirally-even GPDs.*

Although there exists already a huge number of theoretical papers on chirally-even GPDs, only a few attempts have been made to construct models for them (cf. e.g. Ref. [17] and references therein, [18]). They have to reproduce the x-dependence of forward PDFs and the t-dependence known from Nucleon Form Factors. The crucial problem of the correlations between the different GPD variables is still under strong theoretical debate. It remains to be shown to which extent future experiments will be in a position to successfully distinguish between different GPD models. A very first glimpse of the function H can be expected to emerge from mid-term future measurements of various hard exclusive reactions, especially DVCS, at the HERMES experiment upgraded with a Recoil Detector [19] surrounding an *un*polarised target. The projected measurements with an integrated luminosity of 2 fb^{-1} will be able to distinguish between two different sets of 'contemporary' model GPDs over a certain range in the longitudinal momentum variable $\xi = x_B/(2-x_B)$, where x_B is the Bjorken scaling variable. (For details see Refs. [15, 20]).

Considerably less theoretical work and no experimental results exist on either quark or gluon *helicity-flip* GPDs ($H_T^f, \tilde{H}_T^f, E_T^f, \tilde{E}_T^f$). The first proposal how to experimentally access helicity-flip *quark* GPDs is based on the simultaneous production of two ρ mesons [21] at large rapidity separation. This may experimentally be feasible but requires large center-of-mass energies. No model has been proposed yet for these functions and therefore no projections are available yet. Helicity-flip *gluon* GPDs can in principle be accessed through measurements of a distinct angular dependence of the DVCS cross sections [5, 22].

As mentioned above, the forward limit of the second moment of the sum $\frac{1}{2}(H^f + E^f)$ determines the total angular momentum J^f of the parton species f in the nucleus. In this context a forward *orbital angular momentum (OAM) distribution* was introduced [23] as $L^f(x) = J^f(x) - \Delta q_f(x)$, where $J^f(x)$ is the forward limit of the GPD sum $\frac{1}{2}(H^f + E^f)$ and $\Delta q_f(x)$ is the helicity distribution (cf. next chapter). Since the forward limit of the GPD H^f is known from the well-measured PDFs q_f and $\bar{q}_f$, a determination of $J^f(x)$ requires a measurement of the GPD E^f. This is expressed in Fig. 1 by placing the symbol J^f in the outermost ring of the E^f sector. Experimental access to E^f may be achieved through DVCS measurements with an unpolarised beam and a *transversely* polarised target. Projections for the statistical accuracy attainable when measuring the relevant single target-spin asymmetry $A_{UT}(\phi)$ are given in Ref. [29] for both a possible low-luminosity (0.8 fb^{-1}) measurement in the mid-term future and a high-luminosity measurement (100 fb^{-1}) in the far future. Note that the present HERMES measurements using a transversely polarised target, aiming at a determination of u and d quark transversity (cf. next chapter), are planned to collect about 0.15 fb^{-1}. Also, important additional knowledge on E^f can be expected from exclusive vector meson production with transverse target polarisation [17].

4. Parton Distribution Functions

A Parton Distribution as forward limit of a Generalised Parton Distribution depends only on the constituent's momentum fraction x that in deep inelastic lepton-nucleon scattering (DIS) is identified with the Bjorken scaling variable x_B. Additionally, it depends on a resolution scale Q^2, as mentioned above. On the simplest (twist-2) level, as long as multi-parton correlations are not considered, the complete set of quark PDFs consists (for every flavor u, d, s) of number density q_f, *longitudinally* polarised (helicity) distribution Δq_f, and *transversity* distribution δq_f. Only these three functions together form the minimum data set necessary for meaningful confrontations with theoretical models of the ground state of the nucleon.

In-depth studies of DIS over the last two decades resulted in unpolarised quark distributions that are precisely known for valence and well known for sea quarks [24]. Their longitudinally polarised counterparts were measured only recently [25], still with large uncertainties for the sea distributions. In contrast, the *transverse* spin structure of the nucleon is still completely unexplored, even in the forward limit. Several experiments will deliver first data in the near future: COMPASS at CERN [26], HERMES at DESY [27] and RHIC-spin at BNL [28]. Its interpretation will to a large extent be based on the analysis of (single) spin asymmetries in (semi-inclusive) hadron production cross sections. However, precise information on sea quark as well as highly precise data on valence quark distributions can only be expected from the next generation of electron-nucleon scattering experiments. For a brief overview including projections and further references see, e.g., Ref. [29].

For a given flavor, the first moment of the sum (difference) of quark and anti-quark helicity (transversity) distributions represents the axial (tensor) charge of the nucleon. Precise measurements of these integrals are of special interest. The flavor sum of the axial charges, $\Delta\Sigma(Q^2)$, describes the quark's contribution to the total longitudinal spin of the nucleon. Experimentally, a value far below the originally expected non-relativistic limit is found, irrespectively of the underlying renormalisation scheme. This behaviour gave rise to the 'spin crisis' of the nineties. It may be attributed to the fact that the Q^2-evolution of $\Delta\Sigma$ involves the polarised gluon distribution. In contrast, the hitherto totally unmeasured flavor sum of the tensor charges, $\delta\Sigma(Q^2)$, decouples as an all-valence object from gluons and sea quarks and hence is expected to be much closer to the non-relativistic limit (cf. the discussion on lattice QCD results below). Measurements of the tensor charge will give access to the hitherto unmeasured chirally-odd operators in QCD which are of great importance to understand the role of chiral symmetry in the structure of the nucleon [30]. Also, the tensor charge is required as input to calculate the electric dipole moment of the neutron [31] in beyond-the-standard-model theories where the quarks may have electrical dipole moments themselves [32].

Lattice calculations, performed in the context of the operator-product expansion (OPE), lead to reliable results on the 'valence' tensor charge $\delta\Sigma \approx \delta u + \delta d$; their precision improves with time as better methods and computers are used (e.g. 0.562 ± 0.088 in 1997 [33], 0.746 ± 0.047 in 1999 [34]). Nevertheless, these values are far away from the non-relativistic limit of $\frac{5}{3}$ expected for a nucleon consisting of just three valence quarks. In contrast, lattice OPE calculations do not yet lead to reliable results on the 'valence' axial charge $\Delta\Sigma \approx \Delta u + \Delta d$, because the quark-line disconnected diagram cannot be calculated yet. The 'quenched' approximation leads to the value $\Delta\Sigma = 0.18 \pm 0.10$ [33]. This is surprisingly consistent with the earlier result 0.18 ± 0.02 obtained with the simpler method of dynamically staggered fermions [35]. Note that next-to-leading order QCD analyses of experimental data yield $\Delta\Sigma = 0.2 - 0.4$, depending on the factorisation scheme [36]. All numbers quoted above refer to Q^2-values of a few GeV^2.

5. Elastic Form Factors

The t-dependent elastic form factors of the nucleon can in principle be obtained from the Generalised Parton Distributions (once they are known) by integrating over their momentum-fraction variables, i.e. as their first moments. The first moments of the unpolarised chirally-even GPDs H^f and E^f constitute the Dirac (charge) and Pauli (current) form factors $F_1^f(t)$ and $F_2^f(t)$. Note that measurements are usually done for the electric and magnetic Sachs form factors G_E and G_M which are linear combinations of the Dirac and Pauli form factors. The polarised chirally-even GPDs $\tilde{H}^f$ and $\tilde{E}^f$ yield the axial-vector and pseudoscalar form factors g_A^f and h_A^f (also denoted as G_A and G_P). Also for the helicity-flip quark GPDs $(H_T^f, \tilde{H}_T^f, E_T^f, \tilde{E}_T^f)$ form factors exist; for example the first moment of H_T^f at $t = 0$ is the tensor charge of quark species f. The corresponding sectors in Fig. 1 contain no symbols because in the literature there is no unique naming convention nor are there ways for their measurement. For that reason the gluon form factors are also omitted.

Measurements of proton and neutron elastic form factors are accomplished by widely different experimental approaches. Beam energies of a few GeV and less are sufficient to determine their Q^2-dependence which can, in most cases, be approximately described by the $1/Q^2$ dipole form factor. For a recent review of existing measurements and for further references, see Ref. [37].

6. Experimental Requirements

Presently, almost the entire field of GPD-related physics constitutes experimentally virgin territory. Stimulated by the physics insight that can be expected from measurements of GPDs, experimentalists are striving to optimise and upgrade their existing facilities to obtain experimental results that can be used to extract first information about GPDs themselves.

There is little doubt that a completely new step in experimentation, concerning both fixed-target and collider facilities, is required for future GPD measurements. In Fig. 4 a brief summary of the abovementioned major physics topics is given in conjunction with experimental requirements for a future fixed-target electron-nucleon scattering facility. It must offer luminosities of at least 10^{35} per cm^{2}s requiring an accelerator with a duty cycle of 10% or more. Beam energies above 30 GeV are needed to cover a kinematic range suitable for extracting cross sections and their scale dependence in exclusive measurements. For the non-exclusive studies of the hadron structure, including precise measurements of polarised Parton Distribution Functions, variable beam energies in the range of 50-200 GeV are required. In both cases, highest possible polarisation of beam and targets as well as the availability of both beam charges is required to fully exploit the potential of asymmetry measurements. Large-acceptance detector systems with high-rate capabilities must reach a mass resolution of a third of the pion mass, mandatory for the measurement of exclusive channels. For more detailed discussions of the various present-day options for such a facility see, e.g., Refs. [29, 38] which also contain further references. Note that precise measurements of Nucleon Form Factors, while constituting an essential part of the physics menu addressed in this paper, will be pursued at lower-energy facilities.

7. Conclusions

As can be concluded from today's knowledge, major steps in both theory and experimentation are required to accomplish a comprehensive understanding of the angular momentum structure of the nucleon. Physics objectives can be identified for a future fixed-target electron-nucleon scattering facility with polarised targets, high-luminosity, and variable beam energies of 30-200 GeV. Precise measurements of hard exclusive processes will have to be performed in several reaction channels towards a determination of as many as possible Generalised Parton Distributions in a 'global' fit. Independently, very precise measurements of the GPD limiting cases, namely forward Parton Distribution Functions on the one hand, and Nucleon Form Factors on the other, will be of great importance for further developments of the theory, because they eventually will have to complement and confirm the GPD results.

Acknowledgements

I am deeply indebted to M. Diehl and X. Ji for enlightening discussions, to F. Ellinghaus, R. Kaiser and J. Volmer for a careful reading of the manuscript, and to K. Jansen and G. Schierholz for valuable comments on the lattice QCD section. Many thanks to B. Seitz for commenting on, and to A. Hagedorn for producing Fig. 1. The help of K. Pipke in preparing Fig. 4 is acknowledged.

References

1. D. Müller *et al.*, Fortschr. Phys. **42**, 101 (1994).
2. X. Ji, Phys. Rev. Lett. **78**, 610 (1997); Phys. Rev. **D 55**, 7114 (1997).
3. A.V. Radyushkin, Phys. Lett. **B 380**, 417 (1996); Phys. Rev. **D 56**, 5524 (1997).
4. J. Blümlein *et al.*, Nucl. Phys. **B 560**, 283 (1999); Nucl. Phys. **B 581**, 449 (2000).
5. P. Hoodbhoy, X. Ji, Phys. Rev. **D 58**, 054006 (1998).
6. M. Diehl, Eur. Phys. J. **C 19** 485 (2001).
7. X. Artru, M. Mekhfi, Z. Phys. **C 45** 669 (1990).
8. R.L. Jaffe, Phil. Trans. Roy. Soc. Lond. **A 359**, 391 (2001) [arXiv:hep-ph/0008038].
9. A. Freund, Phys. Lett. **B 472**, 412 (2000).
10. A. Freund, M. McDermott, Eur. Phys. J. **C 23**, 651 (2002).
11. A.V. Belitsky, D. Müller, A. Kirchner, Nucl. Phys. **B 629**, 323 (2002).
12. M. Diehl *et al.*, Phys. Lett **B 411**, 193 (1997).
13. HERMES coll., A. Airapetian *et al.*, Phys. Rev. Lett. **87**, 182001 (2001).
14. CLAS coll., S. Stepanyan *et al.*, Phys. Rev. Lett. **87**, 182002 (2001).
15. V.A. Korotkov, W.-D. Nowak, Nucl. Phys. **A 711**, 175 (2002) [arXiv:hep-ph/0207103].
16. F. Ellinghaus for the HERMES coll., Nucl. Phys. **A 711**, 171 (2002) [arXiv:hep-ex/0207029].
17. K. Goeke, M.V. Polyakov, M. Vanderhaeghen, Progr. Part. Nucl. Phys. **47**, 401 (2001).
18. X. Ji, W. Melnitchouk, X. Song, Phys. Rev. **D 56**, 5511 (1997).
19. HERMES coll., DESY PRC 01-01, 2002.
20. V.A. Korotkov, W.-D. Nowak, Eur. Phys. J. **C 23**, 455 (2002).
21. D.Yu. Ivanov *et al.*, CPHT RR 059.0602, arXiv:hep-ph/0209300.
22. A.V. Belitsky, D. Müller, Phys. Lett. **B 486**, 369 (2000).
23. P. Hoodbhoy, X. Ji, W. Lu, Phys. Rev. **D 59**, 014013 (1999).
24. Particle Data Group, D.E. Groom *et al.*, Eur. Phys. J. **C 15**, 1 (2000).
25. M. Beckmann for the HERMES coll., arXiv:hep-ex/0210049, to be publ. in Proc. of the 'Workshop on Testing QCD through Spin Observables in Nuclear Targets', Charlottesville, Virginia/USA, Apr. 18-20, 2002.
26. COMPASS coll., CERN/SPSLC 96-14 (1996).
27. V.A. Korotkov, W.-D. Nowak, K. Oganessyan, Eur. Phys. J. **C 18**, 639 (2001).
28. G. Bunce *et al.*, Ann. Rev. Nucl. Part. Sci. **50**, 525 (2000).
29. W.-D. Nowak, Nucl. Phys. **B** (Proc. Suppl.) **105**, 171 (2002).
30. R.L. Jaffe, MIT-CTP-2685, arXiv:hep-ph/9710465.
31. X. Ji, Proc. of the '15th Int. Spin Symposium', Long Island, New York/USA, Sept. 9-14, 2002.
32. D.A. Demir, M. Pospelov, A. Ritz, arXiv:hep-ph/0208257.
33. S. Aoki *et al.*, Phys. Rev. **D 56**, 433 (1997).
34. S. Capitani *et al.*, Nucl. Phys. **B** (Proc. Suppl.) **79**, 548 (1999).
35. R. Altmeyer *et al.*, Phys. Rev. **D 49**, 3087 (1994).
36. B.W. Filippone, X. Ji, arXiv:hep-ph/0101224.
37. J.J. Kelly, these Proceedings and Proc. of the '15th Int. Spin Symposium', Long Island, New York/USA, Sept. 9-14, 2002.
38. R. Kaiser, Proc. of the '15th Int. Spin Symposium', Long Island, New York/USA, Sept. 9-14, 2002.

Major Physics Topics to Study the Hadron Structure at a High-luminosity Fixed-target eN-facility

Physics	Measured Functions	Processes	Experimental Requests				
			target	σ	t	Q^2	E_{beam}
EXCLUSIVE REACTIONS: • Total quark angular momentum $J^{q\star)}$ $J^q = \sum_f J^f$ (→ Orbital quark ang. mom. L^q) $1^{\underline{st}}$ step: J^u (2006 +)	GPDs: Ji's relation: $J^f = \lim_{t\to 0} \int x dx \left\{ H^f(x,\xi,t) + E^f(x,\xi,t) \right\}$ $1^{\underline{st}}$ extraction: $Im\ H$ (< 2006 ?)	DVCS: $H, E, \tilde{H}, \tilde{E}$ DVEM: pseudo-scalar: $\tilde{H}, \tilde{E}$ vector: H, E	U,T	low σ → high $\int \mathcal{L}$dt	low t → recoil det.	$1 \cdots 20$ GeV^2	$30 \cdots 100$ GeV
SEMI-INCLUSIVE DIS: • PRECISE Measurement of ⟹ tensor charge: (→ chiral symmetry breaking)$^{\star)}$ ⟹ axial charge:	transversity distributions $\delta q_f(x,Q^2) \equiv h_1^f(x,Q^2) \equiv \Delta_T^f(x,Q^2)$ $\delta \sum(Q^2) = \sum_f \int dx \{\delta q_f(x,Q^2) - \delta \bar{q}_f(x,Q^2)\}$ $\Delta \sum(Q^2) = \sum_f \int dx \{\Delta q_f(x,Q^2) + \Delta \bar{q}_f(x,Q^2)\}$	SIDIS + DIS • eN → e' KX for access to $\delta s(x,Q^2)$	T L	high precision →high $\int \mathcal{L}$dt	don't care	$1 \cdots 20$ GeV^2	$50 \cdots 200$ GeV

⋆) fundamental issues !

Summary of requests:

- polarized solid-state targets (T,L);
- sufficient duty cycle (≥ 10 %)
- Variable beam energy (30 ... 200 GeV)

} $\int \mathcal{L}$dt $\geq 100\ fb^{-1}/year$

WDN/KP-Oct.2002

Figure 4. Table of major physics topics in conjunction with experimental requirements.

QUARK DISTRIBUTIONS IN POLARIZED ρ MESON AND ITS COMPARISON WITH THOSE IN PION

A.G. OGANESIAN
INSTITUTE OF THEORETICAL AND EXPERIMENTAL PHYSICS
ITEP, B.CHEREMUSHKINSKAYA 25, MOSCOW,117218,RUSSIA

ABSTRACT

Valence quark distributions in transversally and longitudinally polarizied ρ mesons in the region of intermediate x are obtained by generalizied QCD sum rules. Power corrections up to d=6 are taken into account. Comparison of the results for π and ρ mesons shows, that polarization effects are very significant and SU(6) symmetry of distribution functions is absent. The strong suppression of quark and gluon sea distributions in longitudinally polarizied ρ mesons is found.

INTRODUCTION

Structure functions are one of the most significant characteristics of the inner structure of hadrons . For nucleon and pion there are experimental results (see, e.g. [1]-[3] for the nucleon, and [4,5] , for the pion). For other hadrons, one should use some models based on additional suppositions about inner structure. A very significant question about quark distribution dependence on hadron polarization has no model independent answer (moreover in various models it is usually simply supposed that there is no significant influence). That is why determination of quark distribution functions in a model-independent way in QCD sum rules based only on QCD and the operator product expansion (OPE) seems to be very important task especially for polarized hadron (in this talk we will discuss polarized ρ meson case).

QCD sum rules for valence quark distribution at intermediate x was suggested and developed in [6,7]. The method was based on the fact, that the imaginary part (in s-channel) of 4-point correlator, corresponding to the forward scattering of a virtual photon on the current with the quantum numbers of hadron of interest is dominated by contribution of small distances (at intermediate x) if the virtualities of the photon and hadronic current (q^2 and p^2, respectively) are large and negative $|q^2| \gg |p^2| \gg R_c^{-2}$,

where R_c is the confinement radius. So the operator product expansion in this x region is applicable. Then, comparing dispersion representation of the forward amplitude in terms of physical states with that, calculated in OPE and using Borel transformation to suppress higher resonance contributions, one can find quark distribution functions at intermediate x. Unfortunately, the accuracy of the sum rules for nucleon, obtained in [6], are bad, (especially for d-quark), moreover it was found to be impossible to calculate quark distributions in the π and ρ mesons in this way. The reason is that the sum rules in the form used in [6] have a serious drawback.

E. Steffens and R. Shanidze (eds.), Spin Structure of the Nucleon, 51–59.

This drawback comes from the fact that contribution of nondiagonal transitions wasn't suppressed in sum rules by borelization, and special additional procedure, which was used in [6] to supress them is incorrect for such sum rules. There is no time to discuss this in details, the full analisys can be found in our papers [7] and [8]. In this papers the modified method of calculating of the hadron structure functions (quark distributions in hadrons) was suggested. The problem of supressing the nondiagonal terms is eliminated and valence quark distributions in pion and polarizied ρ meson were calculated.

SECTION 1. THE METHOD

Let me briefly present the method. Consider the non-forward 4-point correlator

$$\Pi(p_1, p_2; q, q') = -i \int d^4x d^4y d^4z \exp[i(p_1 x + qy - p_2 z)] \times$$

$$\langle 0|T\{j^h(x), j^{el}(y), j^{el}(0), j^h(z)|0\rangle \qquad (1)$$

Here p_1 and p_2 are the initial and final momenta carried by hadronic the current j^h, q and $q' = q + p_1 - p_2$ are the initial and final momenta carried by virtual photons and Lorentz indices are omitted. It will be very essential for us to consider unequal p_1 and p_2 and treat p_1 and p_2 as two independent variables. However, we may put $q^2 = q'^2$ and $t = (p_1 - p_2)^2 = 0$. The general form of the double dispersion relation (in p_1^2, p_2^2) of the imaginary part of $\Pi(p_1^2, p_2^2, q^2, s)$ in the s channel after double Borel transformation in p_1^2, p_2^2 has the form

$$B_{M_1^2} B_{M_2^2} \operatorname{Im}\Pi(p_1^2, p_2^2, q^2, s) = \int_0^\infty du_1 \int_0^\infty du_2 \rho(q^2, s, u_1, u_2) \exp\left[-\frac{u_1}{M_1^2} - \frac{u_2}{M_2^2}\right]$$

where M_1^2 and M_2^2 are the squared Borel mass.

One should note, that double Borel transformation eliminates terms with single pole (only on p_1^2 or p_2^2), which correspond to nondiagonal transitions, as was discussed in Introduction. (For simplicity we consider the case of the Lorentz scalar hadronic current. The necessary modifications for the π and ρ mesons will be presented below. The integration region may be divided into four areas:

I. $u_1 < s_0$, $u_2 < s_0$; II. $u_1 < s_0$, $u_2 > s_0$; III. $u_1 > s_0$, $u_2 < s_0$; IV. $u_1 > s_0$, $u_2 > s_0$.

Here s_0 is the continuum threshold in the standard QCD sum rule model of the hadronic spectrum with one lowest resonance plus continuum. Area I obviously corresponds to the resonance contribution and the spectral density in this area can be written as

$$\rho(u_1, u_2, x, Q^2) = g_h^2 \cdot 2\pi F_2(x, Q^2)\delta(u_1 - m_h^2)\delta(u_2 - m_h^2) \qquad (2)$$

where g_h is defined as $\langle 0|j_h|h\rangle = g_h$.

Contribution of the areas II, III, IV, where variables u_1, u_2 are far from the resonance region, are exponentially suppressed and, as usual in the sum rules, the spectral function of a hadron state is described by the bare loop spectral function ρ^0 in the same region.

So, and we would like to emphasise this, we do not need in any additional artificial procedure such as the differentiation with respect to the Borel mass, what is significant advantages of this method. Finally, equating the physical and QCD representations of Π one can write the following sum rules:

$$\text{Im}\,\Pi^0_{QCD} + \text{Power correction} = 2\pi F_2(x, Q^2)g_h^2 \exp\left[-m_h^2(1/M_1^2 + 1/M_2^2)\right]$$

$$\text{Im}\,\Pi^0_{QCD} = \int_0^{s_0} du_1 \int_0^{s_0} du_2 \rho^0(u_1, u_2, x)\exp\left[-(u_1/M_1^2 + u_2/M_2^2)\right] \qquad (3)$$

In what follows, we put $M_1^2 = M_2^2 = 2M^2$. (As was shown in [9], the values of the Borel parameters M_1^2, M_2^2 in the double Borel transformation are about twice as large as those in the ordinary ones).

SECTION 2. QUARK DISTRIBUTIONS IN PION

Let me briefly discuss the main points of the calculation and the results for the pion structure function, which can be treated as a check of the accuracy of the method due to fact that for pion the experimental results are avaible. To find the pion structure function by the method, described in the previous section, one should consider the imaginary part of 4-point correlator (1) with two axial and two electromagnetic currents. Since $\bar{d}(x) = u(x)$, it is enough to find the distribution of the valence u quark in π^+. The most suitable choice of the axial current is $j_5^\mu = \bar{u}\gamma_\mu\gamma_5 d$, and the electromagnetic current is chosen as u-quark current with the unit charge.

The most convenient tensor structure, which is chosen to construct the sum rule, is a structure proportional $P^\mu P^\nu P^\lambda P^\sigma$, where $P = (p_1 + p_2)/2$. The sum rule for the valence u-quark distribution in the pion in the bare loop approximation is [7]:

$$u_\pi(x) = \frac{3}{2\pi^2}\frac{M^2}{f_\pi^2}x(1-x)(1 - \exp[-s_0/M^2])\exp[m_\pi^2/M^2] \qquad (4)$$

In [7] the following corrections to (4) were taken into account:

1. Leading order (LO) perturbative corrections proportional to $\ell n(Q^2/\mu^2)$, where μ^2 is the normalization point. In what follows, the normalization point will be chosen to be equal to the Borel parameter $\mu^2 = M^2$.

2. Power corrections -- higher order terms of OPE. Among them, the dimension-4 correction, proportional to the gluon condensate $\left\langle 0\left|\frac{\alpha_s}{\pi}G_{\mu\nu}^n G_{\mu\nu}^n\right|0\right\rangle$ was first taken into account, but it was found that the gluon condensate contribution to the sum rule vanishes after double borelization. There are two types of vacuum expectation values of dimension 6. One, involving only gluonic fields: $g^3 f^{abc}\left\langle 0\left|G_{\mu\nu}^a G_{\nu\lambda}^b G_{\lambda\nu}^c\right|0\right\rangle$ and the other, proportional to the four-quark operators $a^2 \equiv (2\pi)^4\left\langle 0|\bar{\psi}\psi|0\right\rangle^2$. It was shown in [7] that terms of the first type cancel in the sum rule and only terms of the second type survive. For the latter, one may use the factorization hypothesis which reduces all the terms of

this type to the square of the quark condensate. Finally (see[7]):

$$xu_\pi(x) = \frac{3}{2\pi^2}\frac{M^2}{f_\pi^2}x^2(1-x)\exp[m_\pi^2/M^2]\times$$

$$\left\{\left(1+\frac{\alpha_s(M^2)\ell n(Q_0^2/M^2)}{3\pi}\varphi(x)\right)\left(1-e^{[-s_0/M^2]}\right)-\frac{\alpha_s(M^2)(\alpha_s a^2)}{3^7 2^6 \pi^2 M^6}\frac{\omega(x)}{x^3(1-x)^3}\right\}$$

where $\varphi(x) = \left[\dfrac{1+4x\ell n(1-x)}{x}-\dfrac{2(1-2x)\ell nx}{1-x}\right]$ (5)

$$\omega(x) = (-5784x^4 - 1140x^3 - 20196x^2 + 20628x - 8292)\ell n(2)$$
$$+ 4740x^4 + 8847x^3 + 2066x^2 - 2533x + 1416$$

The function $u_\pi(x)$ may be used as an initial condition at $Q^2 = Q_0^2$ for solution of QCD evolution equations (Dokshitzer-Gribov-Lipatov-Altarelli-Parisi equations). In the numerical calculations (the pion mass is neglected) we choose: the effective $\lambda_{QCD}^{LO} = 200\text{MeV}^2$, $Q_0^2 = 2\text{GeV}^2$, $\alpha_s a^2(1\text{GeV}^2) = 0.13\text{GeV}^2$ [7]. The continuum threshold was varied in the interval $0.8 < s_0 < 1.2\text{GeV}^2$ and it was found, that the results depend only slightly on it's variation. The analysis of the sum rule (5) shows, that it is

xu(x)

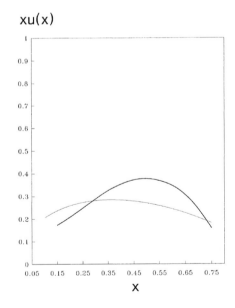

Fig.1. Comparison of the quark distribution in pion (thick line) with experimental fit [4] (thin curve).

X

fulfilled in the region $0.15 < x < 0.7$; the power corrections are less than 30\% and the continuum contribution is small ($< 25\%$). The stability in the Borel mass parameter M^2 dependence in the region $0.4\ \text{GeV}^2 < M^2 < 0.6\ \text{GeV}^2$ is good. The result of our calculation of the valence quark distribution $xu_\pi(x, Q_0^2)$ in the pion is shown in Fig. 1.

In Fig. 1, we plot also the valence u-quark distribution found in [4] by fitting the data on the Drell-Yan process. When comparing with the distribution found here it should be remembered, that the accuracy of our curve is of order of 10 - 20%. The u -quark distribution found from the experiment is also not free from uncertainties (at least 10-20%, see [4,5]). For all these reasons, we consider the agreement of two curves as being good.

Assume, that $u_\pi(x) \sim 1/\sqrt{x}$ at small $x \le 0.15$ according to the Regge behaviour and $u_\pi(x) \sim (1-x)^2$ at large $x \ge 0.7$ according to quark counting rules. Then, matching these functions with (11), one may find the first and the second moments of the u-quark distribution:

$$M_1^\pi = \int_0^1 u_\pi(x)dx \approx 0.84\,(0.85) \; ; \quad M_2^\pi = \int_0^1 xu_\pi(x)dx \approx 0.21(0.23) \qquad (6)$$

where the values in the parentheses correspond to behavior $u_\pi(x) \sim (1-x)$ at large x. The results depend only slightly on the matching points (not more than 5%, when the lower matching point is varied in the region 0.15 - 0.2 and the upper one in the region 0.65 - 0.75). The moment M_1^π has the meaning of the number of u quarks in pion and it should be $M_1^\pi = 1$. Its deviation from 1 characterizes the accuracy of our calculation. The moment M_2^π has the meaning of the pion momentum fraction carried by the valence u quark. Therefore, the valence u and $\overline{d}$ quarks carry about 40% of the total momentum. The valence u-quark distribution in the pion was calculated recently in the instanton model [10]. At intermediate x, the values of $u_\pi(x)$ found in [10] are not more than 20% higher that our results. Our estimation (6) of the second moment of the valence quark distribution can be also compared with the calculation of the second moment of the total (valence plus sea) quark distribution in the pion [11]. The value, obtained in [11], is 0.6 for the total momentum carried by all quarks (valence plus sea) with the accuracy about 10%. Taking into account, that sea quarks usually supposed to carry 15% of the total momentum, one can estimate from the result of [11], that the second moment of one valence quark distribution should be about 20-22%, which is in a good agreement with our result (8). The quark distribution in the pion was calculated also in [12] by using sum rules with nonlocal condensates. Unfortunately it is impossible to perform the comparison directly, because the quark distribution is calculated in [12] only at very low normalization point, but estimations shows, that , our results are in an agreement with those of [12].

SECTION 3. QUARK DISTRIBUTIONS IN ρ MESON

Let us calculate valence u-quark distribution in the ρ^+ meson. The choice of hadronic current in correlator (1) is evident: $j_{(\rho)}^\mu = \overline{u}\gamma_\mu d$. The matrix element is given by $\langle \rho^+ | j_{(\rho)}^\mu | 0 \rangle = (m_\rho^2/g_\rho)e_\mu$, where m_ρ is the ρ meson mass, g_ρ is the $\rho - \gamma$ coupling constant, , $g_\rho^2/4\pi = 1.27$ and e_μ is the ρ meson polarization vector.

It was shown in [8] that in the non-forward amplitude the most suitable tensor structure for determination of u-quark distribution in the longitudinal ρ meson is that proportional to $P^\mu P^\nu P^\lambda P^\sigma$, while u-quark distribution in the transverse ρ can be found by considering the invariant function at the structure $(-P^\mu P^\nu \delta^\sigma_\lambda)$.

In the case of longitudinal ρ meson the tensor structure, that is separated is the same as in the case of the pion. It can be shown, that u-quark distribution in the longitudinal ρ meson can be found from Eq.(8) by substituting $m_\pi \to m_\rho$, $f_\pi \to m_\rho/g_\rho$. Sum rules for $u^L_\rho(x)$ are satisfied in the wide x region: $0.1 < x < 0.85$. The Borel mass M^2 dependence is weak almost in the whole range of x, except $x > 0.7$. Figure. 2 (curve with squares) presents $xu^L_\rho(x)$ as a function of x. The values $M^2 = 1$ GeV^2 and $s_0 = 1.5$ GeV^2, $Q^2_0 = 4$ GeV^2 were chosen, other parameters are the same as in the pion case.

Let us now consider the case of transversally polarizied \rho-meson, i.e., the term proportional to the structure $(-P^\mu P^\nu \delta^\sigma_\lambda)$. The procedure of calculations are the same except two points.

1. In contrast to the pion case, the $\left\langle 0 \left| \frac{\alpha_\pi}{\pi} G^n_{\mu\nu} G^n_{\mu\nu} \right| 0 \right\rangle$ correction for transversally polarized ρ (ρ^T), does not vanish.

2. In contrast to π and ρ^L meson cases, the terms proportional to $g^3 f^{abc} \left\langle 0 \left| G^a_{\mu\nu} G^b_{\nu\lambda} G^c_{\lambda\nu} \right| 0 \right\rangle$ are not cancelled for ρ^T and one should estimate it. But it is not well known; there are only some estimates based on the instanton model [13,14]. The estimation based on the instanton model [13] gives

$$\left\langle fG^3 \right\rangle \equiv g^3 f^{abc} \left\langle 0 \left| G^a_{\mu\nu} G^b_{\nu\lambda} G^c_{\lambda\nu} \right| 0 \right\rangle = -\frac{48\pi^2}{5\rho^2_c} \left\langle 0 \left| (\alpha_s/\pi) G^2_{\mu\nu} \right| 0 \right\rangle \tag{7}$$

where ρ_c is the effective instanton radius. Collecting the results we get finally the following sum rules for the valence u-quark distribution in the transversally polarized ρ meson [8].

$$xu^T_\rho(x) = x \frac{3g^2_\rho}{8\pi^2} \frac{M^4}{m^4_\rho} e^{[m^2_\rho/M^2]} \left\{ \left(1 + \frac{\alpha_s(M^2)\ell n(\frac{Q^2_0}{M^2})}{3\pi} \tau(x) \right) \phi(x) E_1\left(\frac{s_0}{M^2}\right) \right.$$

$$\left. -\frac{\pi^2}{6} \frac{\left\langle 0 \left| (\alpha_s/\pi) G^2 \right| 0 \right\rangle}{M^4 x^2} + \frac{\left\langle fG^3 \right\rangle \xi(x)}{2^8 3^5 M^6 x^3(1-x)^3} + \frac{\alpha_s(M^2)(\alpha_s a^2)}{3^8 2^5 \pi^2 M^6} \frac{\chi(x)}{x^3(1-x)^3} \right\} \tag{8}$$

where $\tau(x) = \left[\frac{4x-1}{\phi(x)} + 4\ell n(1-x) - \frac{2(1-2x+4x^2)\ell nx}{\phi(x)} \right]$;

$\phi(x) = 1 - 2x(1-x)$; $E_1(z) = 1 - (1+z)e^{-z}$;

$\xi(x) = -1639 + 8039\, x - 15233\, x^2 + 10055\, x^3 - 624x^4 - 974x^4$;

$$\chi(x) = 8513 - 41692x + 64589x^2 - 60154x^3 + 99948x^4 - 112516x^5 + 45792x^6$$
$$+ (-180 - 8604x + 53532x^2 - 75492x^3 - 28872x^4 + 109296x^5 - 55440x^6)\ell n2$$

The standard value of the gluonic condensate $\left\langle 0 \left| \frac{\alpha_s}{\pi} G^n_{\mu\nu} G^n_{\mu\nu} \right| 0 \right\rangle = 0.012\,\mathrm{GeV}^2$ was taken in numerical calculations. Equation (11) was used and the effective instanton radius ρ_c was chosen as $\rho_c = 0.5$ fm. This value is between the estimations of [13] ($\rho_c = 1$ fm) and [14] ($\rho_c = 0.33$ fm). (In the recent paper [15] it was argued that the liquid gas instanton model overestimates higher order gluonic condensates and, in order to correct this effect, larger values of ρ_c comparing with [14] should be used). The Borel mass dependence of $xu^T_\rho(x)$ in the interval $0.2 < x < 0.65$ is weak at $0.8 < M^2 < 1.2$ GeV^2. Figure 2 shows $xu^T_\rho(x)$ at $M^2 = 1\mathrm{GeV}^2$ and $Q^2_0 = 4$ GeV^2 (thin curve).

The moments of quark distributions in the longitudinal ρ meson are calculated in the same way, as it was done in for the case of pion: by matching with Regge behavior $u(x) \sim 1/\sqrt{x}$ at low x and with quark counting rule $u(x) \sim (1-x)^2$ at large x. The matching points were chosen as $x = 0.10$ at low x and $x = 0.80$ at large x. The numerical values of the moments for the longitudinally polarized ρ are:

$$M^L_1 \approx 1.06\,(1.05) \; ; \qquad M^L_2 \approx 0.39\,(0.37) \qquad (9)$$

The values of moments, obtained by assuming that $u(x) \sim (1-x)$ at large x are given in the parentheses. Reliable calculation of moments for the case of transversally polarized ρ meson is impossible, because of a narrow applicability domain in x.

SECTION 4. SUMMARY AND DISCUSSION

Fig. 2 gives the comparison of the valence u-quark distributions in the pion and longitudinally and transversally polarized ρ mesons. The shapes of the curves are quite different, especially of $u^T_\rho(x)$) in comparison with $u^L_\rho(x)$ and pion. Strongly different

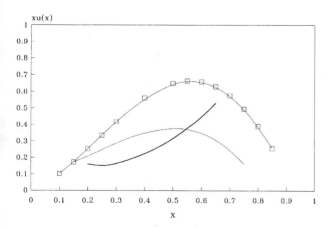

Fig.2. Quark distribution in pion (thin curve), longitudally (curve with squares) and transversally (thick curve) polarized ρ .

fraction, carried by valence quarks and antiquarks in the longitudinal ρ meson is about 0.8, while in the pion it is much less -- about 0.4-0.5. All these differences are many times larger than estimated uncertainties of our results. In the case of u-quark distribution in the pion the main source of them is the value of the quark condensate renorminvariant quantity $\alpha_s a^2$. In our calculations, we took it to be equal to 0.13 GeV^6. In fact, however, it is uncertain by a factor of 2. (Recent determination [16] of this quantity from τ-decay data indicates that it may be two times larger). The perturbative corrections also introduce some uncertainties, especially at large x (x > 0.6) where the LO correction, which is taken into account, is large. The estimation of both effects shows that they may result in 10-20% variation .

For the u-quark distribution in the longitudinally polarized ρ meson, the uncertainties in quark condensate do not play any role, because of higher borel mass values, so the expected accuracy is even better. One should note, that the accuracy of prediction of moments for longitudally polarizied ρ meson are very high, because, as one can see from Fig.2, sum rules predict quark distribution almost in whole region of x (from 0.1 up to 0.9). The fact that the first moment is so close to 1, also confirm that accuracy is high. Large value of the second moment mean, that valence quarks in longitudally polarizied ρ meson carried about 0.8 of the total momentum, so one can conclude that the total (gluon and quark) sea is strongly suppressed. Moreover, this prediction for the second moment is close to those obtained in [17] from quite another sum rules, and also agreed with the lattice calculation results [18], so the sea supression in longitudally polarizied ρ meson can be treated as a theoretically well established fact.

The accuracy of our results for the u-quark distribution in the transversally polarized ρ meson is lower, because of a large role of d=4 and d=6 gluonic condensate contributions. Analysis show (see[8]) that total accuracy is about 30-40% in this case. So one can see, that the difference, obtained in the quark distribution in the pion, longitudinally and transversally polarizied ρ meson in fact are much more larger than any possible uncertainties of the results.

In summary the main physical conclusion are the following:

1. The sea is strongly supressed in the longitudinally polarizied ρ meson.

2. The quark distribution in the polarizied ρ meson are significantly dependend on polarization, and so the quark distributions in the the pion and ρ meson have a not too much in common.

This study was supported in part by Award no. RP2-2247 of U.S. Civilian Research and Development Foundation for Independent States of Former Soviet Union (CRDF), by the Russian Foundation of Basic Research, project no. 00-02-17808 and by INTAS Call 2000, project 587.

REFERENCES

1. H.L.Lai, H.L. et al. (CTEQ Collaboration) (2000) Global QCD analysis of parton srtucture of the nucleon: CTEQ5 parton distribution, Eur. Phys. J. C **12**, 375-392.
2. Martin, A.D., Roberts, R.G., Stirling, W.J. and Thorne, R.S. (1998) Parton distribution: a new global analysis, Eur. Phys. J. C **4**, 463-469.
3. Glück , M., Reya, E. and Vogt, A. (1998) Dynamical parton distribution revisited, Eur. Phys. J. C **5**, 461-470.
4. Glück , M., Reya, E. and Vogt, A. (1992) Pionic parton distributions, Z. Phys. C **53**, 651-656.
5. P.J. Sutton, P.J., Martin, A.D., Roberts, B.G. and Stirling, W.J. (1992) Parton distribution for the pion extracted fromDrell-Yan and promt photon experiments, Phys. Rev. D **45** 2349-2359
6. Belyaev,V.M, and Ioffe, B.L. (1988) Structure functions of deep inelastic lepton-nucleon scattering in QCD (I), Nucl. Phys. B **310**, 548-570.
7. Ioffe, B.L. and Oganesian, A.G. (2000) Improved calculation of quark distributions in hadrons: the case of pion, Eur. Phys. J. C **13**, 485-495.
8. Ioffe, B.L. and Oganesian, A.G. (2001) Valence quark distributions in mesons in generalizied QCD sum rules, Phys. Rev. D **63** 096006.
9. Ioffe, B.L., and Smilga, A.V. (1983) Meson widths and form factors at intermediate momentum transfer in non-perturbative QCD, Nucl. Phys. B**216** 373-407.
10. Dorokhov, A.E. and Tomio, L. (2000) Pion structure function within the unstanton model, Phys. Rev. D **62**, 014016.
11. Belyaev, V.M and Block, B.Yu (1986) Second moment of the pion structure function in QCD, Yad. Fiz. **43**, 706-711.
12. Belitsky, A. (1996) Valence parton density in the pion from QCD sum rules, Phys. Lett. B **386**, 359-369.
13. Shifman, M.A., Vainshtein, A.I.and Zakharov, V.I. (1979) QCD and resonance physics, Nucl. Phys. B **147**, 448-518.
14. Schäfer, T. and Shuryak, E.V (1998), Rev. Mod. Phys. **70**, 323 .
15. Ioffe, B.L and SamsonovA.V. (2000) Correlator of topological charge densities in instanton model in QCD, Phys. At. Nucl. **63**, 1448-1454 .
16. Ioffe, B.L and K.N. Zyablyuk, K.N. (2001) The V-A sum rules and the operator product expansion in complex q^2-plane from τ-decay data, Nucl. Phys. A **687**, 437-453.
17. Oganesian A.and Samsonov, A. (2001) Secomd moment of quark sructure functions in the ρ meson sum rules, JHEP 0109: 002.
18. C. Best, C. et al., (1997) Pion and rho structure functions from lattice QCD, Phys. Rev. D **56**, 2743-2754

NUCLEON ELECTROMAGNETIC FORM FACTORS

JAMES J. KELLY
*Department of Physics, University of Maryland, College Park,
MD 20742*

Abstract. We review data for nucleon electromagnetic form factors, emphasizing recent measurements of G_E/G_M that use recoil or target polarization to minimize systematic errors and model dependence. The data are parametrized in terms of densities that are consistent with the Lorentz contraction of the Breit frame and with pQCD. The dramatic linear decrease in G_{Ep}/G_{Mp} for $1 \leq Q^2 \leq 6$ (GeV/c)^2 demonstrates that the charge is broader than the magnetization of the proton. High precision recoil polarization measurements of G_{En} show clearly the positive core and negative surface charge of the neutron. Combining these measurements, we display spatial densities for u and d quarks in nucleons.

1. Introduction

The electromagnetic structure of structure of nucleons provides fundamental tests of the QCD confinement mechanism, as calculated on the lattice or interpreted with the aid of models. From elastic electron scattering one obtains the Sachs electric and magnetic form factors, which are closely related to the charge and magnetization densities. Dramatic improvements in the quality of these measurements have recently been achieved by using beams that combine high polarization with high intensity and energy together with either polarized targets or measurements of recoil polarization. In this paper we review the current status of nucleon elastic form factors, emphasizing recent polarization measurements, and analyze these data using a model that permits visualization of the underlying charge and magnetization densities.

Matrix elements of the nucleon electromagnetic current operator J^μ take the form

$$\langle N(p', s')|J^\mu|N(p, s)\rangle = \bar{u}(p', s')e\Gamma^\mu u(p, s) \tag{1}$$

E. Steffens and R. Shanidze (eds.), Spin Structure of the Nucleon, 61–74.

where u is a Dirac spinor, p, p' are initial and final momenta, $q = p - p'$ is the momentum transfer, s, s' are spin four-vectors, and where the vertex function

$$\Gamma^\mu = F_1(Q^2)\gamma^\mu + \kappa F_2(Q^2)\frac{i\sigma^{\mu\nu}q_\nu}{2m} \tag{2}$$

features Dirac and Pauli form factors, F_1 and F_2, that depend upon the nucleon structure. Here e is the elementary charge, m is the nucleon mass, κ is the anomalous part of the magnetic moment, and γ^μ and $\sigma^{\mu\nu}$ are the usual Dirac matrices (e.g., [1]). The interpretation of these form factors appears simplest in the nucleon Breit frame where the energy transfer vanishes. In this frame the nucleon approaches with initial momentum $-\vec{q}_B/2$, receives three-momentum transfer $\vec{q}_B$, and leaves with final momentum $\vec{q}_B/2$. Thus, the nucleon Breit frame momentum is defined by $q_B^2 = Q^2 = q^2/(1+\tau)$ where $(\omega, \vec{q})$ is the momentum transfer in the laboratory, $Q^2 = q^2 - \omega^2$ is the spacelike invariant four-momentum transfer, and $\tau = Q^2/4m^2$. In the Breit frame for a particular value of Q^2, the current separates into electric and magnetic contributions [2]

$$\bar{u}(p', s')\Gamma^\mu u(p, s) = \chi_{s'}^\dagger \left(G_E + \frac{i\vec{\sigma} \times \vec{q}_B}{2m} G_M \right) \chi_s \tag{3}$$

where χ_s is a two-component Pauli spinor and where the Sachs form factors are given by

$$G_E = F_1 - \tau\kappa F_2 \qquad G_M = F_1 + \kappa F_2 \tag{4}$$

Early experiments with modest Q^2 suggested that

$$G_{Ep} \approx \frac{G_{Mp}}{\mu_p} \approx \frac{G_{Mn}}{\mu_n} \approx G_D \tag{5}$$

where $G_D(Q^2) = (1 + Q^2/\Lambda^2)^{-2}$ with $\Lambda^2 = 0.71$ $(\text{GeV}/c)^2$ is known as the dipole form factor [3, 4].

2. Form Factors from Polarization Measurements

In the one-photon exchange approximation, the differential cross section for elastic scattering of an electron beam from a stationary nucleon target is given by

$$\frac{d\sigma}{d\Omega} = \frac{\sigma_{NS}}{\epsilon(1+\tau)} \left(\tau G_M^2 + \epsilon G_E^2 \right) \tag{6}$$

where $\epsilon = (1 + (1+\tau)2\tan^2\theta_e/2)^{-1}$ is the transverse polarization of the virtual photon for electron scattering angle θ_e and σ_{NS} is the cross section for a structureless Dirac target. Thus, the traditional Rosenbluth technique separates the electric and magnetic form factors by varying ϵ, but extraction of G_E becomes extremely difficult at large Q^2 because the magnetic contribution becomes increasingly dominant and because it is difficult to control the kinematic variation and radiative corrections with sufficient accuracy when both the form factors and the kinematic coefficients vary rapidly over the acceptance. Alternatively, the electromagnetic ratio

$$g = \frac{G_E}{G_M} = -\sqrt{\frac{\tau(1+\epsilon)}{2\epsilon}} \frac{P'_x}{P'_z} \qquad (7)$$

can be obtained by comparing the components of the nucleon recoil polarization along the momentum transfer direction, denoted by $\hat{z}$, and in the $\hat{x}$ direction transverse to $\hat{z}$ in the scattering plane [5, 6]. For the proton, both components can be measured simultaneously using a polarimeter in the focal plane of a magnetic spectrometer, thereby minimizing systematic uncertainties due to beam polarization, analyzing power, and kinematic parameters. The systematic uncertainty due to precession of the proton spin in the magnetic spectrometer is usually much smaller than the uncertainties in comparing the cross sections obtained with different kinematical conditions and acceptances needed for the Rosenbluth method.

Figure 1 compares recent high-precision measurements of the proton electromagnetic ratio performed at Jefferson Laboratory (JLab) [7, 8] with earlier Rosenbluth data obtained at SLAC [9, 10]. I had expected the new JLab experiments to confirm the SLAC findings that $G_{Ep} \approx G_{Mp}/\mu_p$, merely improving their precision, and privately regarded these experiments as preparation for more challenging measurements of the neutron charge form factor via recoil polarization. Surprisingly, however, we found the strong linear decrease shown in Fig. 1 instead. The systematic uncertainties, primarily due to spin precession in the magnetic spectrometer, are shown by the hatched region and were substantially reduced in the second experiment at larger Q^2 where the deviation from unity is strongest. A recent re-analysis of the SLAC data failed to discover any systematic correction which could account for this disagreement [11] and I am unaware of any plausible mechanism which could produce a failure of the one-photon approximation at this level. Hopefully, a recent *super Rosenbluth* experiment [12] designed to minimize systematic uncertainties will help clarify this discrepancy. Until those results become available, I will rely on the recoil polarization measurements and discard the cross section data for $Q^2 > 1$ $(GeV/c)^2$ with larger, and possibly seriously underestimated, systematic uncertainties.

For the neutron one must use quasifree scattering from a neutron in a nuclear target and correct for the effects of Fermi motion, meson-exchange currents, and final-state interactions also. Both recoil and target polarization for quasifree knockout give similar PWIA formulas for the form factor

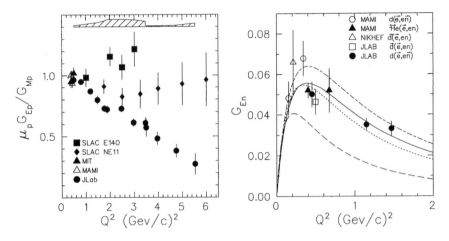

Figure 1. Recent G_E/G_M data for the proton (left) and neutron (right). The hatched region indicates the systematic uncertainty in JLab recoil polarization measurements for the proton. The proton data are from: SLAC E140 [9], SLAC NE11 [10], MIT [13], MAMI [14], and JLab [7, 8]. The neutron data are from: MAMI $d(\vec{e}, e'\vec{n})$ [15], MAMI $^3\vec{\text{He}}(\vec{e}, e'n)$ [16, 17], NIKHEF $\vec{d}(\vec{e}, e'n)$ [18], JLab $\vec{d}(\vec{e}, e'n)$ [19], and JLab $d(\vec{e}, e'\vec{n})$ [20]. See text for explanation of G_{En} curves.

ratio, but the systematic errors and the nuclear physics corrections are appreciably different. Therefore, confident extraction of G_{En}/G_{Mn} benefits from comparison of data for both recoil and target polarization. Data from recent experiments of these types are also shown in Figure 1 and additional data from Mainz on $d(\vec{e}, e'\vec{n})$ at $Q^2 = 0.6$ and 0.8 and from JLab on $\vec{d}(\vec{e}, e'n)$ at $Q^2 = 1.0$ (GeV/c)^2 are expected soon. Details of the high-precision JLab experiment on $d(\vec{e}, e'\vec{n})$ at $Q^2 = 0.45$, 1.15, and 1.47 (GeV/c)^2 may be found in Ref. [20]; corrections for nuclear physics and acceptance averaging have not yet been applied but are expected to be small. Several two-parameter fits based upon the Galster parametrization are shown also. Our fit (solid) to these data and additional data from Refs. [21, 22] remains rather close to the original Galster fit (dotted) to electron-deuteron elastic scattering data for $Q^2 < 0.7$ (GeV/c)^2 [23] despite the rather larger model dependence of the Rosenbluth method. The highest curve shows a fit by Schmieden [24] to a subset of the polarization data for $Q^2 < 0.7$ (GeV/c)^2. The dashed line shows the commonly quoted Platchkov analysis of more recent elastic scattering data (not shown) that used the Paris potential, but other realistic interactions produce variations greater than the spread between the highest and lowest curves in this figure [25]; thus, the model dependence of the Rosenbluth method is at least this large while the model dependence of the recoil polarization method for $Q^2 > 0.5$ (GeV/c)^2 is less than 10% [26]. Therefore, polarization techniques provide much more accurate measurements of G_{En} than the Rosenbluth method and show that G_{En} for $Q^2 < 1.5$ $(\text{GeV}/c)^2$ is substantially larger than the common Platchkov parametrization, although it remains compatible with the Galster parameterization.

Polarization measurements of nucleon electromagnetic ratios and selected cross section data for the form factors relative to G_D are compared in Figs. 2-3 with representative calculations. The chiral soliton model of Holzwarth (dotted lines) predicted the linear behavior of G_{Ep}/G_{Mp} but fails to reproduce neutron form factors [27, 28]. The light-cone diquark model (long dashes) needs only 5 parameters to obtain a reasonable fit for modest Q^2 [29], but except for G_{En} its form factors fall too rapidly for $Q^2 > 1$ $(\text{GeV}/c)^2$. The point-form spectator approximation (PFSA) using pointlike constituent quarks and a Goldstone boson exchange interaction fitted to spectroscopic data successfully describes a wider range of Q^2 (short dashes) without fitting additional parameters to the form factors [30]. Finally, a light-front calculation using one-gluon exchange and constituent-quark form factors fitted to $Q^2 < 1$ $(\text{GeV}/c)^2$ provides a good fit (dash-dot) up to about 4 $(\text{GeV}/c)^2$ [31]. However, none of the available theoretical calculations provides a truly quantitative description for all four form factors over a wide range of Q^2. The differences between these models are largest for G_{En}, which is especially sensitive to small mixed-symmetry and deformed components of the nucleon wave function. Clearly it will be very important to extend the G_{En} data to larger Q^2 — a proposal to measure ${}^3\vec{He}(\vec{e}, e'n)$ up to 3.4 $(\text{GeV}/c)^2$ has been approved at JLab [32] and proposals for higher Q^2 are under development.

Figures 2-3 also show fits made by Lomon [33] using an extension of the Gari-Krümpelmann model [34] that interpolates between vector meson dominance (VMD) at low Q^2 and perturbative QCD (pQCD) at high Q^2. This type of parametrization is very useful for nuclear physics calculations, but offers no insight into the spatial distributions of charge and magnetization within nucleons. In the next section we offer an alternative phenomenology in terms of spatial densities.

3. Fitted Densities

3.1. RELATIVISTIC INVERSION

Although rigorous comparisons between theory and experiment must be made at the level of form factors, for many it would seem desirable to extract charge and magnetization densities from the corresponding form factors because our intuition is usually stronger in space than in momentum transfer. Intrinsic charge and magnetic form factors, $\tilde{\rho}_{ch}(k)$ and $\tilde{\rho}_m(k)$, may

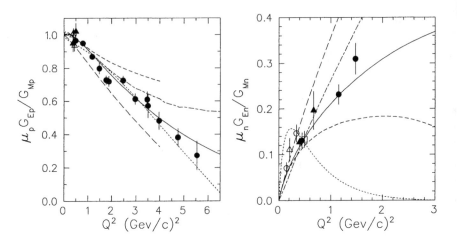

Figure 2. Polarization data for electromagnetic ratios are compared with representative calculations: chiral soliton (dotted) [27, 28], light-cone diquark (long dashes) [29], PFSA (short dashes) [30], light-front OGE with constituent form factors (dash-dot) [31]. The data have the same legend as Fig. 1.

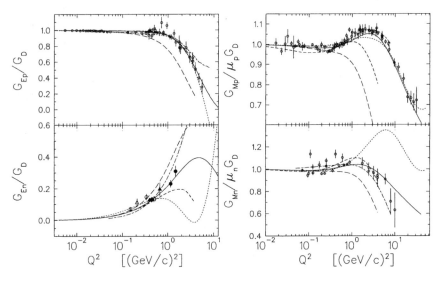

Figure 3. Selected form factor data are compared with representative calculations: chiral soliton (dotted) [27, 28], light-cone diquark (long dashes) [29], PFSA (short dashes) [30], light-front OGE with constituent form factors (dash-dot) [31].

be defined in terms of the Sachs form factors by

$$\tilde{\rho}_{ch}(k) = G_E(Q^2)(1+\tau)^{\lambda_E}$$
$$\mu\tilde{\rho}_m(k) = G_M(Q^2)(1+\tau)^{\lambda_M} \tag{8}$$

where the intrinsic spatial frequency k is related to the invariant momentum transfer Q by the Breit-frame boost

$$k^2 = \frac{Q^2}{1+\tau} \tag{9}$$

and where the model-dependent exponents, λ_E and λ_M will be discussed shortly. Due to the Lorentz contraction of spatial distributions in the Breit frame, a measurement with Breit-frame momentum transfer $q_B = Q$ probes a reduced spatial frequency k in the rest frame. In fact, the intrinsic frequencies accessible to elastic scattering with spacelike momentum transfer are limited to $k < 2m$ such that the asymptotic Sachs form factors in the limit $Q^2 \to \infty$ are determined by the intrinsic form factors in the immediate vicinity of the limiting frequency $k_m = 2m$. This limitation can be understood as a consequence of relativistic position fluctuations, known as of *zitterbewegung*, that smooth out radial variations on scales smaller than the Compton wavelength.

Using a quark cluster model Licht and Pagnamenta [35] originally proposed to use $\lambda_E = \lambda_M = 1$, but these choices do not conform with pQCD scaling unless one imposes upon both form factors the somewhat artificial constraint $\tilde{\rho}(k_m) = 0$. Mitra and Kumari [36] then demonstrated that a more symmetric version of the quark cluster model that is also applicable to inelastic transitions suggests $\lambda_E = \lambda_M = 2$ and is compatible with pQCD scaling without constraining $\tilde{\rho}(k_m)$. By contrast, the exponents obtained by Ji [37] using the chiral soliton model, $\lambda_E = 0$ and $\lambda_M = 1$, would require the even more restrictive conditions $\tilde{\rho}_{ch}(k_m) = \tilde{\rho}'_{ch}(k_m) = \tilde{\rho}_m(k_m) = 0$ to obtain pQCD scaling and will not be considered further here. For the present work we employ $\lambda_E = \lambda_M = 2$ and refer to Ref. [38] for a more comprehensive analysis.

Having proposed a model for the intrinsic form factors that is consistent with both the Lorentz contraction of the Breit frame and pQCD scaling, we can obtain spatial densities by simply inverting the Fourier transforms according to

$$\rho(r) = \frac{2}{\pi} \int_0^\infty dk \, k^2 j_0(kr)\tilde{\rho}(k) \tag{10}$$

Note that we have normalized these densities such that

$$\int_0^\infty dr \, r^2 \rho_{ch}(r) = Z \qquad \int_0^\infty dr \, r^2 \rho_m(r) = 1 \tag{11}$$

where $Z = 0, 1$ is the nucleon charge. Unfortunately, we cannot claim that $\rho_{ch}(r)$ and $\rho_m(r)$ are the true charge and magnetization densities in the nucleon rest frame because the boost operator for a composite system depends upon the interactions among its constituents. Nevertheless, this model can be used to fit the form factor data using an intuitively appealing spatial representation that is consistent with relativity and with pQCD.

To illustrate the effects of the relativistic transformation between density and form factor, suppose that the density has a Gaussian shape

$$\rho(r) = \frac{4}{b^3\sqrt{\pi}}\exp\left(-(r/b)^2\right) \qquad \tilde{\rho}(k) = \exp\left(-(kb/2)^2\right) \qquad (12)$$

that is typical of quark models. The form factor obtained using Eq. (8) with $\lambda = 2$ is compared with the familiar dipole form factor in Fig. 4 for several choices of b; note with $b = 0.556$ fm the Gaussian parametrization has the same rms radius as the dipole form factor. These curves display the same general features as the data for G_{Ep}, G_{Mp} and G_{Mn}: for low Q^2 the form factor is close to the dipole form while for large Q^2 one finds an asymptotic limit for G/G_D that depends sensitively upon b but is less than unity for reasonable values. The greatest sensitivity to the shape of the density is found in a transition region for Q^2 that ranges from several tenths to several $(\text{GeV}/c)^2$, depending upon b. Thus, data with similar general features can be fit by modulating a basic Gaussian with an even polynomial, where the polynomial degree can be minimized by an optimal choice of b. For G_{En} one need only require the polynomial part of $\tilde{\rho}(k)$ to begin with k^2 to ensure that the net charge vanishes.

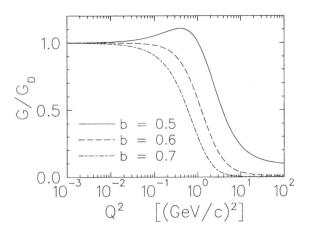

Figure 4. The ratio between Sachs form factor with $\lambda = 2$ and the dipole form factor is shown for a Gaussian intrinsic density using several values of the oscillator parameter, b, listed with units in fm.

3.2. FITTING PROCEDURES

The model dependence of the fitted form factor can be minimized by expanding the density in a complete set of radial basis functions, such that

$$\rho(r) = \sum_n a_n f_n(r) \Longrightarrow \tilde{\rho}(k) = \sum_n a_n \tilde{f}_n(k) \qquad (13)$$

where

$$\tilde{f}_n(k) = \int_0^\infty dr\, r^2 j_0(kr) f_n(r) \qquad (14)$$

represents basis functions in momentum space. The expansion coefficients, a_n, are fitted to form factor data subject to several minimally restrictive constraints to be discussed shortly. Analyses of this type are often described as model independent because a complete basis can reproduce any physically reasonable density; if a sufficient number of terms are included in the fitting procedure the dependence of the fitted density upon the assumptions of the model is minimized. By contrast, simple parametrizations like the Galster model severely constrain the shape of the fitted density.

A convenient choice of basis is the Laguerre-Gaussian expansion (LGE)

$$\begin{aligned} f_n(r) &= e^{-x^2} L_n^{1/2}(2x^2) \\ \tilde{f}_n(k) &= \frac{\sqrt{\pi}}{4} b^3 (-)^n e^{-y^2} L_n^{1/2}(2y^2) \end{aligned} \qquad (15)$$

where $x = r/b$, $y = kb/2$, and L_n^a is a generalized Laguerre polynomial. A significant advantage of the LGE is that the number of terms needed to provide a reasonable approximation to the density can be minimized by choosing b in accordance with the natural radial scale. We chose $b = 0.556$ fm such that the mean-square radius of the Gaussian factor is consistent with that of the common dipole parametrization of Sachs form factors. We then find that the magnitude of a_n decreases rapidly with n, but the quality of the fit and the shape of the density are actually independent of b over a wide range. An arbitrarily large number of terms can be included by using a penalty function to constrain high-frequency contributions with an envelope of the form

$$k > k_{max} \Longrightarrow |\tilde{\rho}(k)| < |\tilde{\rho}(k_{max})| (k_{max}/k)^4 \qquad (16)$$

where k_{max} is the largest frequency for which experimental data are available. This condition ensures that fitted density does not have an unphysical cusp at the origin but does permit the density sufficient flexibility to estimate the uncertainty due to the absence of data for $k > k_{max}$. In addition, one constrains the density for very large radii. Details of these procedures can be found in Ref. [38] and references cited therein.

3.3. RESULTS

We fit all four nucleon electromagnetic form factors using a data selection that emphasizes recent polarization methods where available. For G_{Mp} and G_{Mn} we employed the highest quality cross section data in each range of Q^2. For G_{Ep} we used the recoil polarization data from Refs. [7, 8, 14, 13] and chose cross section data from Refs. [39, 40] for low Q^2 but omitted the higher Q^2 Rosenbluth data from Refs. [9, 10]. For G_{En} we use recoil and target polarization data corrected for nuclear physics effects and use the results of an analysis of the deutron quadrupole form factor by Schiavilla and Sick [21]; Rosenbluth data for elastic or quasielastic scattering from deuterium were omitted. We also include the measurement of $\langle r^2 \rangle_n$ by Kopecky et al. [22] using the energy dependence for the transmission of thermal neutrons through liquid ^{208}Pb. A more complete review of these selections and omissions can be found in Ref. [38].

Figure 5 shows fits to the form factor data using the LGE model with $\lambda = 2$. Where data are available the widths of the error bands are governed by the statistical quality of data while for large Q^2 the growth of these uncertainties is limited by the large-k constraint specified by Eq. (16). These fits are generally very good, but in the G_{Mn} data there remain appreciable systematic differences between data sets that probably reflect errors in the efficiency calibration for some of the experiments.

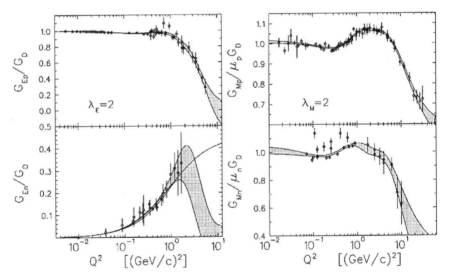

Figure 5.   LGE fits to selected data for nucleon form factors using $\lambda_E = \lambda_M = 2$. For G_{En} we also show a Galster fit.

Figure 6 compares the four fitted densities. For the proton we find that the charge is distributed over a larger volume than the magnetization. The difference between rms radii is not large, 0.883(14) for charge versus 0.851(26) fm for magnetization, but the large difference in interior densities reflects the strong decrease in G_{Ep}/G_{Mp} for $Q^2 > 1$ (GeV/c)^2. The magnetization density for the neutron is very similar to that for the proton, but closer examination shows that its distribution is slightly wider. For the purposes of comparing shapes, the neutron charge density is shown scaled to the interior magnetization. Despite the limited range of Q^2 and larger uncertainties in the G_{En} data, the neutron charge density is determined with useful precision.

The neutron charge density features a positive interior and negative surface. In the meson-baryon picture these characteristics are explained in terms of quantum fluctuations of the type $n \leftrightarrow p\pi^-$ in which the light negative meson is found at larger radius than the heavier positive core. Alternatively, in the quark model these features arise from incomplete cancellation between u and d quark distributions that are similar but not identical in shape. Using a symmetric two-flavor quark model of the nucleon charge densities

$$\rho_p(r) = \frac{4}{3}u(r) - \frac{1}{3}d(r) \qquad \rho_n(r) = -\frac{2}{3}u(r) + \frac{2}{3}d(r) \qquad (17)$$

one can obtain the quark densities

$$u(r) = \rho_p(r) + \frac{1}{2}\rho_n(r) \qquad d(r) = \rho_p(r) + 2\rho_n(r) \qquad (18)$$

where $u(r)$ is the radial distribution for an up quark in the proton or a down quark in the neutron while $d(r)$ is the distribution for a down quark in the proton or an up quark in the neutron. These quark densities are displayed in Fig. 7 weighted by r^2 to emphasize the surface region. We find that the u distribution is slightly broader than the d distribution, which is consistent with the repulsive color hyperfine interaction between like quarks needed to explain the $N - \Delta$ mass splitting. The slightly negative $d(r)$ near 1 fm suggests a $\bar{d}$ contribution from the pion cloud. The secondary lobes near 1.4 fm appear to be robust features of the data — elimination of these features seriously degrades fits to data for $Q^2 \sim 1$ (GeV/c)^2 — and might arise from mixed symmetry or $\ell = 2$ admixtures with larger radii than the dominant S-state configuration.

4. Conclusions

The advent of highly polarized electron beams with large currents and high energy coupled with advances in recoil polarimetry and polarized targets permit much more precise measurements of the nucleon electromagnetic form factor ratio, G_E/G_M, than was possible with the Rosenbluth

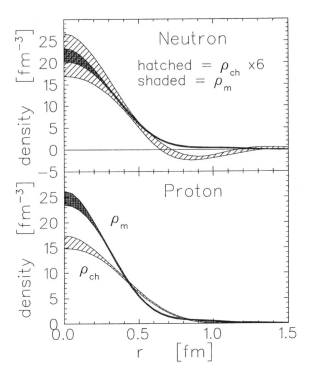

Figure 6. Nucleon electromagnetic densities using $\lambda_E = \lambda_M = 2$.

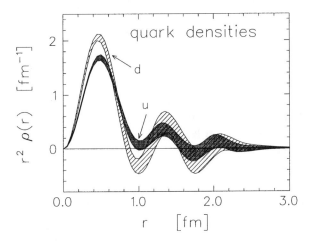

Figure 7. Quark densities using $\lambda_E = \lambda_M = 2$.

technique at large Q^2. Despite the fact that Rosenbluth data suggested $\mu_p G_{Ep}/G_{Mp} \approx 1$ for $Q^2 < 6$ $(\text{GeV}/c)^2$, recoil polarization measurements at Jefferson Laboratory show a strong, nearly linear, decrease for $Q^2 > 1$

$(\mathrm{GeV}/c)^2$ that demonstrates that the charge density is significantly broader than the magnetization density of the proton. Similarly, recoil and target polarization measurements show that G_{En} is substantially larger than the commonly quoted Platchkov analysis of deuteron elastic scattering using the Paris potential yet remains surprisingly close to the original Galster parametrization despite the prohibitively large model dependence of Rosenbluth separations for G_{En}.

We have developed a phenomenological model of these form factors in terms of spatial densities that is consistent with pQCD at large Q^2 and with the Lorentz contraction of the Breit frame relative to the rest frame. The model dependence of the fitted densities is minimized by using an expansion in a complete set of basis functions with minimally restrictive constraints upon the behavior for either large frequency or large radius. The flexibility of the fitted form factor for frequencies beyond the measured momentum transfer provides an estimate of the incompleteness error in the extracted density. The error envelopes for the magnetization densities and the proton charge density are quite narrow, but the uncertainty in the neutron charge density is significantly larger because the data are still limited to $Q^2 < 1.5$ $(\mathrm{GeV}/c)^2$ and are not as precise as for the other form factors. Nevertheless, the precision is already quite useful and will improve when the next generation of $^3\vec{H}e(\vec{e},e'n)$ experiments reaches about 3.4 $(\mathrm{GeV}/c)^2$ [32].

We find that the neutron and proton magnetizations densities are quite similar but that the proton charge density is significantly broader. The neutron charge density results from incomplete cancellation between u and d quark densities with slightly different shapes that leaves a positive core surrounded by negative surface charge. By comparing the fitted neutron and proton charge densities, we find that the distribution of like quarks in the nucleon is broader than the distribution of the unlike quark.

Acknowledgements

I thank D. Quing, S. Simula, M. Radici, and F. Coester for tabulated calculations. The support of the U.S. National Science Foundation under grant PHY-9971819 is gratefully acknowledged.

References

1. J. D. Bjorken and S. D. Drell, *Relativistic Quantum Mechanics* (McGraw-Hill, New York, 1964).
2. R. G. Sachs, Phys. Rev. **126**, 2256 (1962).
3. E. B. Hughes, T. A. Griffey, M. R. Yearian, and R. Hofstadter, Phys. Rev. **139**, B458 (1965).

4.   J. J. R. Dunning, K. W. Chen, A. A. Cone, G. Hartwig, N. F. Ramsey, J. K. Walker, and R. Wilson, Phys. Rev. **141**, 1286 (1966).
5.   N. Dombey, Rev. Mod. Phys. **41**, 236 (1969).
6.   R. G. Arnold, C. E. Carlson, and F. Gross, Phys. Rev. **C 23**, 363 (1981).
7.   M. K. Jones et al., Phys. Rev. Lett. **84**, 1398 (2000).
8.   O. Gayou et al., Phys. Rev. Lett. **88**, 092301 (2002).
9.   R. C. Walker et al., Phys. Rev. **D 49**, 5671 (1994).
10. L. Andivahis et al., Phys. Rev. **D 50**, 5491 (1994).
11. J. Arrington, arXiv:nucl-ex/0205019 (unpublished).
12. J. Arrington et al., Jefferson Laboratory proposal E01-001, 2001.
13. B. D. Milbrath et al., Phys. Rev. Lett. **82**, 2221 (1999).
14. T. Pospischil et al., Eur. Phys. J. A **12**, 125 (2001).
15. C. Herberg et al., Eur. Phys. J. A **5**, 131 (1999).
16. J. Golak, G. Ziemer, H. Kamada, H. Witala, and W. Glöckle, Phys. Rev. **C 63**, 034006 (2001).
17. D. Rohe et al., Phys. Rev. Lett. **83**, 4257 (1999).
18. I. Passchier et al., Phys. Rev. Lett. **82**, 4988 (1999).
19. H. Zhu et al., Phys. Rev. Lett. **87**, 081801 (2001).
20. R. Madey et al., to be published in Proceedings of Electron Nucleus Scattering VII by Eur. Phys. J. A (unpublished).
21. R. Schiavilla and I. Sick, Phys. Rev. **C 64**, 041002(R) (2001).
22. S. Kopecky, J. A. Harvey, N. W. Hill, M. Krenn, M. Pernicka, P. Riehs, and S. Steiner, Phys. Rev. **C 56**, 2220 (1997).
23. S. Galster, H. Klein, J. Moritz, K. Schmidt, D. Wegener, and J. Bleckwenn, Nucl. Phys. **B32**, 221 (1971).
24. H. Schmieden, in *Proceedings of the 8th International Conference on the Structure of Baryons*, edited by D. W. Menze and B. Metsch (World Scientific, Singapore, 1999), pp. 356–367.
25. S. Platchkov et al., Nucl. Phys. **A510**, 740 (1990).
26. H. Arenhövel, Phys. Lett. **B199**, 13 (1987).
27. G. Holzwarth, Zeit. Phys. A **356**, 339 (1996).
28. G. Holzwarth, arXiv:hep-ph/0201138 (unpublished).
29. B.-Q. Ma, D. Qing, and I. Schmidt, Phys. Rev. **C 65**, 035205 (2002).
30. R. Wagenbrunn, S. Boffi, W. Klink, W. Plessas, and M. Radici, Phys. Lett. **B 511**, 33 (2001).
31. S. Simula, arXiv:nucl-th/0105024 (unpublished).
32. G. Cates et al., Jefferson Laboratory proposal E02-013, 2002.
33. E. L. Lomon, arXiv:nucl-th/0203081 (unpublished).
34. M. F. Gari and W. Krümpelmann, Zeit. Phys. A **322**, 689 (1985).
35. A. L. Licht and A. Pagnamenta, Phys. Rev. **D 2**, 1156 (1970).
36. A. N. Mitra and I. Kumari, Phys. Rev. **D 15**, 261 (1977).
37. X. Ji, Phys. Lett. **B254**, 456 (1991).
38. J. J. Kelly, arXiv:hep-ph/0204239 (unpublished).
39. G. G. Simon, C. Schmitt, F. Borkowski, and V. H. Walther, Nucl. Phys. **A333**, 381 (1980).
40. L. E. Price, J. R. Dunning, M. Goitein, K. Hanson, T. Kirk, and R. Wilson, Phys. Rev. **D 4**, 45 (1971).

HADRON PRODUCTION AND LAMBDA POLARIZATION IN DIS

ARAM KOTZINIAN
Yerevan Physics Institute
Alikhanian Brothers St. 2, Yerevan AM-375036, Armenia
and JINR
Moscow region, RU-141980 Dubna, Russia

Abstract

The role of hadronization mechanism in polarization phenomena in DIS and a purity method for extraction of polarized distribution functions are discussed. A model for the longitudinal polarization of Λ^0 baryons produced in deep-inelastic lepton scattering is proposed. Within the context of our model, the NOMAD data imply that the intrinsic strangeness associated with a valence quark has anticorrelated polarization. We also compare our model predictions with the data from the HERMES experiment. Predictions of our model for the COMPASS experiment are also given.

1. Introduction

It is generally accepted that information from deep inelastic scattering (DIS) experiments is an excellent source for investigating the internal structure of nucleon. Experimental progress in recent years allows to investigate the semi inclusive DIS (SIDIS). There is hope that, for example, the measurement of different hadron production asymmetries on proton and neutron targets will allow a further flavor separation of polarized quark distributions.

The knowledge of the hadronization mechanism is playing a very important role in the interpretation of SIDIS data. Traditionally one distinguishes two regions for hadron production: the current fragmentation region: $x_F > 0$ and the target fragmentation region: $x_F < 0$. The common assumption is that when selecting hadrons in the current fragmentation

E. Steffens and R. Shanidze (eds.), Spin Structure of the Nucleon, 75–83.

region and imposing a cut $z > 0.2$ we are dealing with the quark fragmentation.

The measurements of the longitudinal polarization of Λ^0 hyperon produced in SIDIS was believed providing two type of information. In the target fragmentation region it will provide an access to the polarization of intrinsic strangeness of the nucleon [2]. And in the current fragmentation region it will measure the polarization transfer from the quark q to Λ hyperon [3, 4, 5]: $C_q^\Lambda(z) \equiv \Delta D_q^\Lambda(z)/D_q^\Lambda(z)$, where $D_q^\Lambda(z)$ and $\Delta D_q^\Lambda(z)$ are unpolarized and polarized fragmentation functions. Several experimental measurements of Λ^0 polarization have been made in neutrino and antineutrino DIS. Longitudinal polarization of Λ^0 hyperons was first observed in the old bubble chamber (anti) neutrino experiments [6, 7, 8]. The NO-MAD Collaboration has recently published new and interesting results on Λ^0 and $\bar{\Lambda}^0$ polarization with much larger statistics [9]. There are also recent results on longitudinal polarization of Λ^0 hyperons from polarized charged lepton nucleon DIS processes coming from E665 [10] and HERMES [11] experiments.

In Sec. 2 the purity method for polarized distribution function extraction is discussed. A method of calculation of the longitudinal polarization of Λ^0 hyperons produced in SIDIS is presented in Sec. 3, and our model predictions are compared to the available data in Sec. 4. Finally, in Sec. 5 some conclusions are presented.

2.   Remarks on the purity method

To make flavor decomposition into polarized quark distributions the purity method has been used in the HERMES analysis [12]. In the LO approximation the virtual photon asymmetry is given by

$$A_1^h \simeq \frac{\sum_q e_q^2 \Delta q(x, Q^2) \int_{z_{min}}^1 dz D_q^h(z, Q^2)}{\sum_q e_q^2 q(x, Q^2) \int_{z_{min}}^1 dz D_q^h(z, Q^2)}. \tag{1}$$

This equation can be rewritten in the form

$$A_1^h \simeq \sum_q P_q^h(x) \frac{\Delta q(x)}{q(x)}, \tag{2}$$

where the purity, $P_q^h(x)$, is defined as

$$P_q^h(x) = \frac{e_q^2 q(x) \int_{z_{min}}^1 dz D_q^h(z)}{\sum_{q'} e_{q'}^2 q'(x) \int_{z_{min}}^1 dz D_{q'}^h(z)}, \tag{3}$$

and calculated using an unpolarized Monte-Carlo event generator LEPTO [13] – JETSET [14]. Then using measured asymmetries for different hadrons one

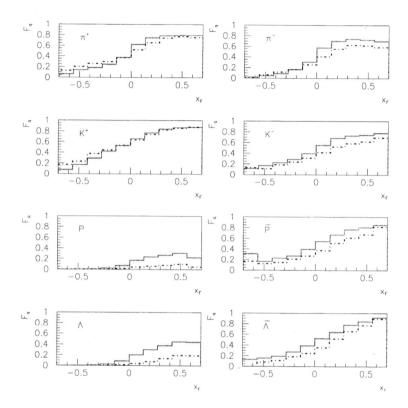

Figure 1.   Fraction of hadrons originating from quark fragmentation for COMPASS (solid line) and HERMES energies (dot-dashed line)

can find $\Delta q(x)$ by solving Eq. 2. The main assumption of this method is that all hadrons in the current fragmentation region with $z > 0.2$ are produced from the quark fragmentation so there is no additional terms in both the numerator and the denominator of Eq. 1. However, this assumption fails for moderate energies in the LUND fragmentation model incorporated in the JETSET program. In this program there is a pointer which shows the origin of produced hadrons. In Fig. 1, the fraction of events with hadrons produced via quark fragmentation

$$F_q = \frac{N_{hadron}(from\ quark\ fragmenation)}{N_{hadron}(tot)} \tag{4}$$

is presented for different hadrons as a function of x_F. As one can see this fraction is less than one even at large values of x_F. Thus the assumption that hadrons in the current fragmentation region are produced only via quark fragmentation is not valid in the LUND model and purities obtained with LEPTO Monte Carlo generator include contributions from the target remnant fragmentation.

When one takes into account the contribution from the target remnant Eq. 1 for virtual photon asymmetry is modified:

$$A_1^h \simeq \frac{\sum_q \Delta q(x, Q^2) D_q^h(z, Q^2) + \Delta M^{h/p}(x, z, Q^2)}{\sum_q q(x, Q^2) D_q^h(z, Q^2) + M^{h/p}(x, z, Q^2)}. \tag{5}$$

The additional contributions from diquark fragmentation and other sources arise in numerator and denominator. Then it is easy to see that the purity method can be applied only if these terms are proportional to the corresponding quark distribution functions with the same proportionality coefficient:

$$\frac{\Delta q(x, Q^2)}{q(x, Q^2)} = \frac{\Delta M^{h/p}(x, z, Q^2)}{M^{h/p}(x, z, Q^2)}. \tag{6}$$

It is not evident that this relation holds for all quark and hadron types. Taking this into account it is hard to trust the polarized quark distribution obtained by the purity method.

3. Λ^0 production and polarization in DIS

Here a short description of the approach and results of the work [1] will be given.

Strange hadrons can be produced in SIDIS due to the struck quark or the nucleon remnant diquark fragmentation. The longitudinal polarization of the lepton can be transfered to strange hadrons during this fragmentation process. Λ^0 hyperons can be produced *promptly* or as a decay product of heavier strange baryons (Σ^0, Ξ, $\Sigma^\star$). Therefore to predict a polarization for Λ^0 hyperons in a given kinematic domain one needs to know the relative yields of Λ^0's produced in different channels and their polarization. We take into account all these effects by explicitly tracing the Λ^0 origin predicted by the fragmentation model and assigning the polarization according to the polarized intrinsic strangeness model in the diquark fragmentation and by SU(6) and Burhardt-Jaffe models for the quark fragmentation.

3.1. POLARIZED INTRINSIC STRANGENESS MODEL

The main idea of the polarized intrinsic strangeness model applied to semi-inclusive DIS is that the polarization of s quarks and $\bar{s}$ antiquarks in the hidden strangeness component of the nucleon wave function should be (anti)correlated with that of the struck quark. This correlation is described

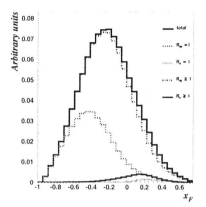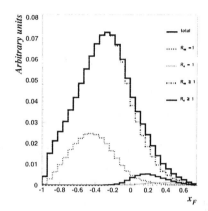

Figure 2.   Predictions for the x_F distributions of all Λ^0 hyperons (solid line), of those originating from diquark fragmentation and of those originating from quark fragmentation, for the two model variants A and B, as explained in the legend on the plots. The left panel is for ν_μ CC DIS with $E_\nu = 43.8$ GeV, and the right panel for μ^+ DIS with $E_\mu = 160$ GeV.

by the spin correlation coefficients C_{sq}: $P_s = C_{sq}P_q$, where P_q and P_s are the polarizations of the initial struck (anti)quark and remnant s quark. In principle, C_{sq} can be different for the valence and sea quarks. We leave $C_{sq_{val}}$ and $C_{sq_{sea}}$ as free parameters, that are fixed in a fit to the NOMAD data [9].

3.2. POLARIZATION OF STRANGE HADRONS IN (DI)QUARK FRAGMENTATION

We define the quantization axis along the three-momentum vector of the exchanged boson. To calculate the polarization of Λ^0 hyperons produced in the diquark fragmentation we assume the combination of a non-relativistic $SU(6)$ quark-diquark wave function and the polarized intrinsic strangeness model described above. The polarization of Λ^0 hyperons produced in the quark fragmentation via a strange baryon (Y) is calculated as: $P^q_{\Lambda^0}(Y) = -C_q^{\Lambda^0}(Y)P_q$, where $C_q^{\Lambda^0}(Y)$ is the corresponding spin transfer coefficient, P_q is the struck quark polarization which depends on the process. We use $SU(6)$ and BJ models to compute $C_q^{\Lambda^0}(Y)$.

3.3. FRAGMENTATION MODEL

To describe Λ^0 production and polarization in the full x_F interval, we use the LUND string fragmentation model, as incorporated into the JETSET7.4 program. We use the LEPTO6.5.1 Monte Carlo event generator to simulate charged-lepton and (anti)neutrino DIS processes. We introduce two rank counters: R_{qq} and R_q which correspond to the particle rank from the di-quark and quark ends of the string, correspondingly. A hadron with $R_{qq} = 1$ or $R_q = 1$ would contain the diquark or the quark from one of the ends of the string. However, one should perhaps not rely too heavily on the tag-ging specified in the LUND model. Therefore, we consider the following two variant fragmentation models:

Model A: The hyperon contains the stuck quark (the remnant diquark) only if $R_q = 1$ ($R_{qq} = 1$).

Model B: The hyperon contains the stuck quark (the remnant diquark) if $R_q \geq 1$ and $R_{qq} \neq 1$ ($R_{qq} \geq 1$ and $R_q \neq 1$).

Clearly, Model B weakens the Lund tagging criterion by averaging over the string, while retaining information on the end of the string where the hadron originated.

In the framework of JETSET, it is possible to trace the particles' parent-age. We use this information to check the origins of the strange hyperons produced in different kinematic domains, especially at various x_F. Accord-ing to the LEPTO and JETSET event generators, the x_F distribution of the diquark to Λ^0 fragmentation is weighted towards large negative x_F.

However, its tail in the $x_F > 0$ region overwhelms the quark to Λ^0 x_F distribution at these beam energies. In Fig. 2, we show the x_F distributions of Λ^0 hyperons produced in diquark and quark fragmentation, as well as the final x_F distributions. These distributions are shown for ν_μ CC DIS at the NOMAD mean neutrino energy $E_\nu = 43.8$ GeV, and for μ^+ DIS at the COMPASS muon beam energy $E_\mu = 160$ GeV. The relatively small fraction of the Λ^0 hyperons produced by quark fragmentation in the region $x_F > 0$ is related to the relatively small centre-of-mass energies - about 3.6 GeV for HERMES, about 4.5 GeV for NOMAD, about 8.7 GeV for COMPASS, and about 15 GeV for the E665 experiment - which correspond to low W.

We vary the two correlation coefficients $C_{sq_{val}}$ and $C_{sq_{sea}}$ in fitting Mod-els A and B to the following 4 NOMAD points:

1) νp: $P_x^\Lambda = -0.26 \pm 0.05(stat)$, 2) νn: $P_x^\Lambda = -0.09 \pm 0.04(stat)$, 3) $W^2 < 15$ GeV^2: $P_x^\Lambda(W^2 < 15) = -0.34 \pm 0.06(stat)$, 4) $W^2 > 15$ GeV^2: $P_x^\Lambda(W^2 > 15) = -0.06 \pm 0.04(stat)$. We find from these fits similar values for both the $SU(6)$ and BJ models: $C_{sq_{val}} = -0.35 \pm 0.05$, $C_{sq_{sea}} = -0.95 \pm 0.05$ (model A) and $C_{sq_{val}} = -0.25 \pm 0.05$, $C_{sq_{sea}} = 0.15 \pm 0.05$ (model B).

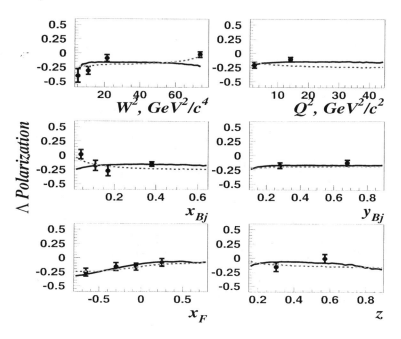

Figure 3.   The predictions of model A - solid line and model B - dashed line, for the polarization of Λ hyperons produced in ν_μ charged-current DIS interactions off nuclei as functions of W^2, Q^2, x_{Bj}, y_{Bj}, x_F and z (at $x_F > 0$). The points with error bars are from NOMAD experiment .

4. Results on Λ polarization

In Figs. 3 and 4 we show our model predictions compared to the available data from the NOMAD [9] and HERMES [11] experiments. One can conclude that our model quite well describes all the data. The NOMAD Collaboration has measured separately the polarization of Λ^0 hyperons produced off proton and neutron targets. We observe good agreement, within the statistical errors, between the model B description and the NOMAD data while model A quite well reproducing the polarization of Λ^0 hyperons produced from an isoscalar target fails to describe target nucleon effects. We provide many possibilities for further checks of our approach for future data (see for details [1]).

The COMPASS Collaboration plans to investigate the polarization of Λ^0 hyperons produced in the DIS of polarized μ^+ on a ^6LiD target. The beam energy and polarization are 160 GeV and -0.8, respectively. Thanks to the large statistics expected in this experiment, one can select kinematic

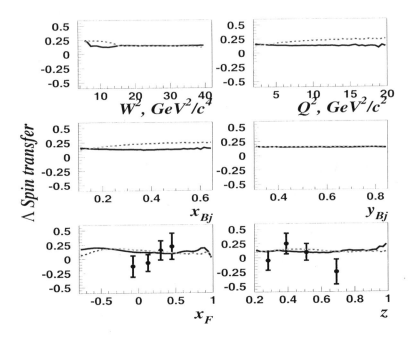

Figure 4. The predictions of model A - solid line, model B - dashed line, for the spin transfer to Λ hyperons produced in e^+ DIS interactions off nuclei as functions of W^2, Q^2, x_{Bj}, y_{Bj}, x_F and z (at $x_F > 0$). We assume $E_e = 27.5$ GeV, and the points with error bars are from HERMES experiment.

regions where the predicted polarization is very sensitive to the value of the spin correlation coefficient for sea quarks, $C_{sq_{sea}}$. For example, in the region $x_F > -0.2$, which is experimentally accessible, and imposing the cut $0.5 < y < 0.9$, one ensure a large spin transfer from the incident lepton to the struck quark, and enhance the contribution from the sea quarks. The predicted Λ polarization is presented in Table 1.

TABLE 1.   Predicted Λ polarization for COMPASS experiment

P_Λ (%)	Target nucleon		
	isoscalar	proton	neutron
model A	-7.3	-7.3	-7.2
model B	-0.4	-0.4	-0.4

As one can see two models gives quite different predictions and the new measurement of the Λ^0 polarization can give preference to the one of described models.

5. Conclusions

To treat the polarization phenomena in SIDIS it is very important to trace the origin of hadrons. The modern Monte Carlo event generators are very successful in the description of unpolarized SIDIS. At the same time as we have learned in Sec. 2 essential part of hadrons are not produced by the quark fragmentation even in the current fragmentation region.

As we have demonstrated in Sec. 3 and Sec. 4 one can successfully describe the existing data on Λ^0 longitudinal polarization in the combined SU(6) and intrinsic strangeness model when one take into account the origin of the strange hyperons.

In contrast, the purity method, also based on the LEPTO event generator, assumes that all hadrons in the current fragmentation region are produced via quark fragmentation. As one can see in Fig. 1 this is not a good approximation. Thus, results obtained by this method are highly questionable.

References

1.   J. Ellis, A. Kotzinian, D .Naumov, hep-ph/0204206
2.   J. Ellis, M. Karliner, D. E. Kharzeev and M. G. Sapozhnikov, Phys. Lett. **B353** (1995) 3129; Nucl. Phys. **A673** 256 (2000);
     J. Ellis, D. E. Kharzeev and A. Kotzinian, Z. Phys. **C69** 467 (1996).
3.   The COMPASS Proposal, CERN/SPSLC 96-14, SPSC/P297, March 1996.
4.   A. Kotzinian, A. Bravar and D. von Harrach, Eur. Phys. J. **C2** 329 (1998).
5.   M. Burkardt and R. L. Jaffe, Phys. Rev. Lett. **70** 2537 (1993) R. L. Jaffe, Phys. Rev. **D54**, R6581 (1996).
6.   J. T. Jones et al., [WA21 Collaboration], Z. Phys. **C28** 23 (1987).
7.   S. Willocq et al., [WA59 Collaboration], Z. Phys. **C53** 207 (1992).
8.   D. DeProspo et al., [E632 Collaboration], Phys. Rev. **D50** 6691 (1994).
9.   P. Astier et al., [NOMAD Collaboration], Nucl. Phys. **B588** (2000) 3;
     D. V. Naumov [NOMAD Collaboration], AIP Conf. Proc. **570** (2001) 489, hep-ph/0101325;
     P. Astier et al. [NOMAD Collaboration], Nucl. Phys. **B605**, 3 (2001) hep-ex/0103047.
10. M. R. Adams et al. [E665 Collaboration], Eur. Phys. J. **C17**, 263 (2000), hep-ex/9911004.
11. A. Airapetian et al. [HERMES Collaboration], Phys. Rev. **B64**, 112005 (2001) hep-ex/9911017
12. H. E. Jackson, at this workshope; hep-ex/0208015
13. G. Ingelman, A. Edin and J. Rathsman, Comp. Phys. Commun. **101** 108 (1997).
14. T. Sjöstrand, *PYTHIA 5.7 and JETSET 7.4: physics and manual*, LU-TP-95-20 (1995), hep-ph/9508391;
     T. Sjöstrand, Comp. Phys. Commun. **39** 347 (1986), **43** 367 (1987).

ELECTROPRODUCTION OF REAL PHOTONS VIA DVCS AT HERMES

M. AMARIAN

Yerevan Physics Institute, Alikhanian Brs. st. 2,
Yerevan, 375036, Armenia
and
DESY Zeuthen, 15738 Zeuthen, Germany
E-mail: amarian@mail.desy.de
FOR THE HERMES COLLABORATION

Abstract. We present experimental results on single beam-spin asymmetry and lepton-charge asymmetry in hard exclusive electroproduction of real photons on hydrogen target obtained by HERMES Collaboration using E=27.5 GeV polarized electron/positron beam of HERA at DESY.

1. Introduction

Inclusive deep-inelastic lepton-nucleon scattering experiments played a significant role in our understanding of the internal structure of the nucleon. Recent progress in this field is related to exclusive reactions and the *Generalized Parton Distributions (GPD's)* which take into account the dynamical correlations between partons of different momenta. The GPD-framework can be used to evaluate a wide range of observables, such as electromagnetic form factors, inclusive parton distributions and exclusive meson-production cross sections.

One of the cleanest channels to probe GPD's is Deeply Virtual Compton scattering (DVCS) i.e. the observation of a multi-GeV photon radiated from a single quark in deep-inelastic lepton-nucleon scattering , when highly virtual photon (with large Q^2) is absorbed and the squared four-momentum transfer between initial- and final state nucleon, t, is small. This reaction is unique, as the produced real photon carries direct information about partonic structure of the nucleon without being distorted by hadronization processes like in meson production.

E. Steffens and R. Shanidze (eds.), Spin Structure of the Nucleon, 85–94.

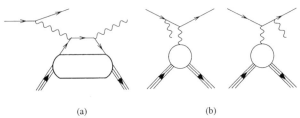

(a)   (b)

Figure 1. Feynman diagramm for: a)- DVCS and b)- BH subprocesses in the electro-production of real photons

Another particular interest to DVCS measurements was triggered by its connection to the spin structure of the nucleon. In [1] it has been argued that DVCS provides information on the total angular momentum $(J = 1/2(\Delta\Sigma) + L)$ of the partons in the nucleon. In view of the non-trivial spin structure of the nucleon, independent information on the angular momentum decomposition of the nucleon spin is highly desirable. It should be noted, though, that the link between DVCS measurements and spin decomposition of the nucleon will require significant experimental and theoretical efforts. There is a flow of papers devoted to DVCS and exclusive reactions establishing formalizm of GPD's and providing theoretical tools for their interpretaion [2], [3], [4]

Although virtual compton scattering has been discussed for decades experimental information on this process until recently was practically absent.

2. General Formalizm

Experimental challenge for the electroproduction of photons is due to intensive background when incoming and/or scattered positron radiate via Bethe-Heithler (BH) subprocess. Diagrammatically both: the radiation from the struck quark and the lepton lines are depicted in fig.1.

In order to separate these subprocesses one has to either select a kinematical domain where the BH radiation is suppressed or to use the interference between the BH and DVCS. In the last case one gets an access to the amplitudes rather than cross sections, which provides unique opportunity to measure imaginary and real parts of DVCS amplitudes for the first time.

In the present experiment we exploit the latter option. Following [5] the cross section for leptoproduction of real photons can be written as

$$\frac{d\sigma}{d\phi dt dQ^2 dx_{Bj}} = \frac{x_{Bj} y^2}{32 \, (2\pi^4) \, Q^4} \frac{|\tau_{BH} + \tau_{DVCS}|^2}{\left(1 + 4x_{Bj}^2 m^2/Q^2\right)^{1/2}}, \qquad (1)$$

where τ_{BH}, τ_{DVCS} are the BH and DVCS amplitudes, $y = \nu/E_e$ is the fraction of the incoming lepton energy carried by the virtual photon, Q^2 its four-momentum squared, and m the proton mass.

The full expression with all terms included is very lenghty, therefore it is omitted here. However interference term in polarized case can be expressed as follows

$$(\tau_{BH}^* \tau_{DVCS} + \tau_{DVCS}^* \tau_{BH}) \sim \tag{2}$$

$$e_l P_l \left[-\sin\phi \cdot \sqrt{\frac{1+\epsilon}{\epsilon}} \mathrm{Im}\tilde{M}^{1,1} + \sin 2\phi \cdot \mathrm{Im}\tilde{M}^{0,1} \right]$$

For the unpolarized case interference term looks as follows

$$\tau_{BH}^* \tau_{DVCS} + \tau_{DVCS}^* \tau_{BH} \sim$$

$$e_l \cdot \frac{e^6}{t} \frac{m}{Q} \cdot \frac{4\sqrt{2}}{x_{Bj}} \cdot \frac{1}{\sqrt{1-x_{Bj}}} \left[\cos(\phi) \cdot \frac{1}{\sqrt{\epsilon}(1-\epsilon)} \mathrm{Re}\tilde{M}^{1,1} \right.$$

$$-(2\phi) \cdot \sqrt{\frac{1+\epsilon}{1-on}} \mathrm{Re}\tilde{M}^{0,1}$$

$$\left. -\cos(3\phi) \cdot \sqrt{\frac{\epsilon}{1-\epsilon}} \mathrm{Re}\tilde{M}^{-1,1} \right] + O\left(1/Q^2\right) \tag{3}$$

In this equations the quantities $\tilde{M}^{\lambda,\lambda'}$ are linear combinations of the DVCS helicity amplitudes $M_{h,h'}^{\lambda,\lambda'}$ with λ, λ' (h, h') representing the helicities of the initial and final-state photon (target nucleon)

$$\tilde{M}^{\lambda,\lambda'} =$$

$$\Delta_T \cdot \left[(1 - x_{Bj}) G_M - (1 - x_b/2) F_2 \right] \cdot M_{-1/2,-1/2}^{\lambda,\lambda'}$$

$$+ \frac{\Delta_T}{m} \cdot \left[G_M - (1 - x_b/2) F_2 \right] \cdot M_{1/2,1/2}^{\lambda,\lambda'}$$

$$+ \left[x_{Bj}^2 G_M + \frac{\Delta_T^2}{2m^2} F_2 \right] \cdot M_{1/2,-1/2}^{\lambda,\lambda'} - \frac{\Delta_T^2}{2m^2} F_2 \cdot M_{-1/2,1/2}^{\lambda,\lambda'} \tag{4}$$

the variable $\Delta_T^2 = -t$ is the absolute value of four-momentum transfer between virtual and real photons. The quantities G_M and F_2 represent Pauli and Dirac form-factors of the nucleon.

The coefficient $e_l = \pm 1$ (lepton charge) and $P_l = \pm 1$ (lepton helicity) of the incident lepton show the sensitivity of the interference term to the difference between electron and positron scattering and spin direction of incoming beam respectively. Azimuthal angle ϕ is defined as the angle between the lepton scattering plane and the plane defined by the virtual and real photon. The dependence of the cross section on ϕ gives rise to azimuthal asymmetries. The kinematical quantity ϵ represents the longitudinal-transverse polarization of the virtual photon.

Since $|\tau_{DVCS}|^2$ is very small and $|\tau_{BH}|^2$ does not depend on the lepton beam helicity, the interference term given above will dominate measurements of the cross-section asymmetry with respect to the lepton beam helicity P_l. From the equation (2) above it can be seen that such measurements provide information on the imaginary part of the DVCS amplitudes, while from equation (3) one can see that using lepton-charge asymmetry, i.e. electron beam versus positron beam, thanks to HERA, real part of the amplitudes are accessible.

3. Experimental Results

3.1. BEAM-SPIN ASYMMETRY

In order to extract imaginary part of DVCS amplitudes HERMES experiment [6] used E=27.5 GeV polarized positron beam of HERA ring and unpolarized hydrogen target data. Only events with single photon cluster and scattered positron have been selected for DVCS analysis. To assure hard scale in the process cuts $Q^2 \geq 1.0 GeV^2$ and invariant energy $W^2 \geq 4.0 GeV^2$ have been applied. The constraint on $\nu \leq 24 GeV$ comes from 3.5 GeV energy threshold of the trigger. In order to select exclusive events additional cut on missing mass $M_x = (q + P_p - P_\gamma)$, where q, P_p and P_γ are four-momenta of virtual-γ, target nucleon and produced photon respectively, to be in the range of $-1.5 \leq M_x \leq 1.7$ GeV has been used.

In fig.2 measured missing mass squared distribution is presented together with Monte Carlo (MC) simulation of real-γ event rate normalized to the same number of scattered deep inelastic positrons with above mentioned cuts. Upper panel shows the full kinematical range of our measurement, while the lower panel zooms exclusive region. Experimental data are presented as points and the shaded histograms represent results of MC simulation. The light shaded part reflects the contribution of real photons from fragmentation processes and the gaussian type darker hatched histgram represents exclusive elastic nucleon to nucleon (N-N) transition. The little tiny band on top of all histograms at low missing masses represents N-Δ transition yield. As one can see, althogh missing mass resolution is smeared due to calorimeter resolution, nevertheless contribution from N-Δ subprocess is suppressed by order of magnitude due to the dominance of elastic scattering in BH at very low Q^2.

The M_x^2 spectrum is possibly contaminated by photon pairs from π^0 decay that enter the same calorimeter segment and are misidentified as

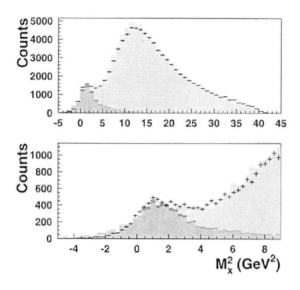

Figure 2.   Missing mass distribution in the electroproduction of real photons

one photon. It may also happen that one of the π^0 decay photons escapes detection. First of all we have experimentally determined the rate of reconstructed π^0 events and then using Monte-Carlo simulation estimated background of single γ events from π^0 decay. It was found that π^0 mesons produced as fragmentation products in deep-inelastic scattering may contaminate the exclusive part of the photon spectrum by 6% and the contamination due to exclusive π^0 production was found to be 2.5%. Taken together the total contamination is estimated to be 8.5%.

The DVCS-BH interference terms can be extracted from the dependence of the data on the azimuthal angle ϕ. In order to have an almost full ϕ-coverage, events were selected with $15 < \theta_{\gamma\gamma^*} < 70$ mrad, where $\theta_{\gamma\gamma^*}$ represents the angle between the directions of the virtual photon and the real photon. A MC-simulation shows that for angles smaller than 15 mrad with present algorithm, the granularity of the calorimeter (9×9 cm^2) is insufficient to reliably determine the angle ϕ, (it was shown later that with improved position recnstruction the cut on the minimal value of $\theta_{\gamma\gamma^*}$ could be as low as 2 mrad). For angles larger than 70 mrad, the ϕ-acceptance is restricted. The average ϕ-resolution in the selected $\theta_{\gamma\gamma^*}$ range is about 0.14 rad.

In fig.3 the ϕ distribution of single beam-spin asymmetry A_{LU} defined as

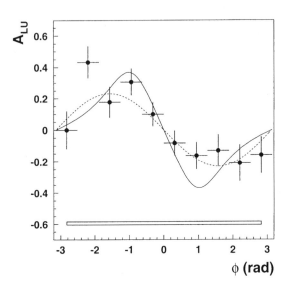

Figure 3. Azimuthal angle distribution of single beam-spin asymmetry A_{LU}

$$A_{LU}(\phi) = \frac{1}{\langle |P_l| \rangle} \cdot \frac{N^+(\phi) - N^-(\phi)}{N^+(\phi) + N^-(\phi))}, \tag{5}$$

is plotted [7] together with simple $\sin\phi$ fit and more complex theory prediction [8]. Here N^+ and N^- represent the luminosity-normalized yields of events with corresponding beam helicity states, $\langle |P_l| \rangle$ is the average magnitude of the beam polarization, and the subscripts L and U denote a longitudinally polarized beam and an unpolarized target. The data displayed in Fig. ?? have been selected requiring a missing mass between -1.5 and +1.7 GeV, i.e. -3σ below and $+1\sigma$ above $M_x = m$, and represent 4015 events. An asymmetric M_x-range was chosen to minimize the influence of the DIS-fragmentation background while optimizing the statistics. Inteference between two processes can occure only and if only final states for both of them are identical. As we have demonstrated that BH process in the kinematical range of present experiment is elastic and has only a few per cent contribution from higher states, we conclude that interference term has the same small contribution from non-exclusive γ production.

In order to see how beam-spin asymmetry depends on missing mass it

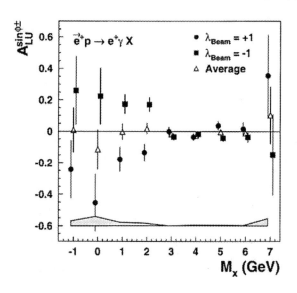

Figure 4. The $\sin\phi$-moment $A_{LU}^{\sin\phi^{\pm}}$ as a function of the missing mass for positive beam helicity (circles), negative beam helicity (squares) and the averaged helicity (open triangles). A negative value is assigned to M_x if $M_x^2 < 0$. The error bars are statistical only. The systematic uncertainty is represented by the error band at the bottom of the figure.

is useful to form $\sin\phi$ weighted moments

$$A_{LU}^{\sin\phi^{\pm}} = \frac{2}{N^{\pm}} \sum_{i=1}^{N^{\pm}} \frac{\sin\phi_i}{|P_l|_i}, \qquad (6)$$

where the superscript $\pm$ refers to the helicity of the positron beam. In fig.4 the extracted values of $A_{LU}^{\sin\phi^{\pm}}$ are plotted versus the missing mass M_x for the two helicity states λ_{Beam} of the positron beam.

As one can see the sign of the $\sin\phi$-moment is opposite for the two beam helicities, in agreement with the expectations for the helicity dependence of the relevant DVCS-BH interference term. The spin averaged data can be used to estimate an upper limit of a possible false asymmetry which — averaged for M_x between -1.5 and +1.7 GeV — amounts to -0.03 ± 0.04. The data for two different beam helicities contain the same information and therefore can be combined to evaluate beam-spin analyzing power $A_{LU}^{\sin\phi}$:

$$A_{LU}^{\sin\phi} = \frac{2}{N} \sum_{i=1}^{N} \frac{\sin\phi_i}{(P_l)_i}, \qquad (7)$$

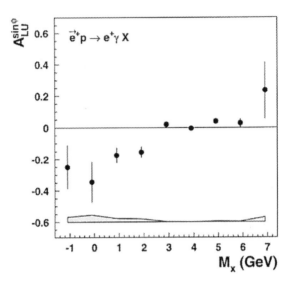

Figure 5. The beam-spin analyzing power $A_{LU}^{\sin\phi}$ for hard electroproduction of photons on hydrogen as a function of the missing mass. The systematic uncertainty is represented by the error band at the bottom of the figure.

where $N = N^+ + N^-$. In contrast to Eq. 6, the sign of the beam polarization is explicitly taken into account, thus distinguishing the two helicity states.

By combining the $A_{LU}^{\sin\phi}$ data in the same M_x region as was used for fig.3 ($-1.5 < M_x < 1.7$ GeV), an average value of -0.23 $\pm$ 0.04 (stat) $\pm$ 0.03 (syst) is obtained. The average values of the kinematic variables corresponding to this measurement are: $\langle x \rangle = 0.11$, $\langle Q^2 \rangle = 2.6$ GeV^2 and $\langle -t \rangle = 0.27$ GeV^2.

3.2. BEAM-CHARGE ASYMMETRY

As it was discussed in previous section one can access the real part of DVCS amplitudes by measuring beam-charge asymmetry. For this purpose HERMES used unpolarized (spin-averaged) electron and positron scattering data on unpolarized hydrogen target citeellinghaus. In this case according to general formalizm azimuthal angle distribution of the A^{ch} presented in fig.6 for the leading amplitude should have cosϕ behavior. A^{ch} is defined as

$$A^{ch}(\phi) = \frac{N^+(\phi) - N^-(\phi)}{N^+(\phi) + N^-(\phi)}, \qquad (8)$$

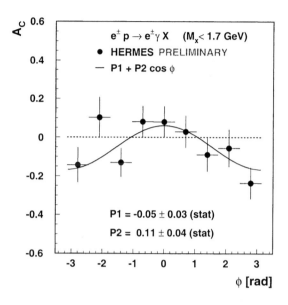

Figure 6. Beam-charge asymmetry A_{ch} versus azimuthal angle for missing mass range below 1.7 GeV. The curve is the result of simple fit with constant term plus cosϕ.

where N^+ and N^- represent the luminosity-normalized yields of events for positron and electron beam respectively.

By analogy with sin-moments, in fig.7 the cosϕ-moment of lepton charge asymmetry defined as

$$A^{ch}_{\cos \phi} = \sum_{i=1}^{N} \frac{\cos \phi_i}{N_i}, \qquad (9)$$

is plotted versus missing mass.

As one can see there is significant beam-charge asymmetry at low values of missing masses, while at higher masses it is consistent with zero. The measure value of the cosϕ amplitude extracted from the fit on fig.6 is found to be $P2 = 0.11 \pm 0.04(stat.)$.

4. Conclusions

We have presented experimental results on beam-spin and lepton-charge asymmetry in exclusive electroproduction of real photons. Due to the interference between BH and DVCS subprocess these asymmetries allow for the first time to get an access to imaginary and real parts of DVCS amplitudes. Further studies of DVCS at HERMES are related to the beam-spin asymmetry on nuclear targets, target-spin asymmetry on proton and

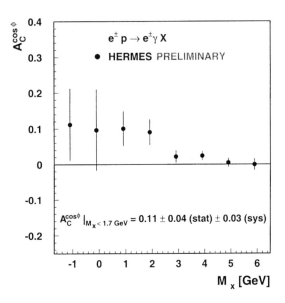

Figure 7. The cosϕ-moment of beam-charge asymmetry versus missing mass

deuteron targets, lepton charge asymmetry on deuteron, higher sin$n\phi$ and cos$n\phi$ moments, kinematical distributions of all observables. Significant improvement on experimental side is expected to take place after installation of recoil detector to measure final state nucleon.

5. Acknowledgments

I would like to thank organizers and the staff of Yerevan Physics Institute for such an interesting workshop and very pleasant atmosphere.

References

1. X. Ji, Phys. Rev. Lett. 78 (1997) 610.
2. F-M. Dittes et al., Phys. Lett. B 209 (1988) 325.
3. D. Muller et al., Fortsch. Phys. 42 (1994) 101.
4. A.V. Radyushkin, Phys. Lett. B 385 (1996) 333.
5. M. Diehl et al., Phys. Lett. B 411 (1997) 193.
6. HERMES Collaboration, K. Ackerstaff et al., Nucl. Instr. Meth. A 417 (1998) 230.
7. HERMES Collaboration, A. Airapetian et al., Phys. Rev. Lett. 87: 182001, 2001.
8. N. Kivel, M. Polyakov and M. Vanderhaeghen, Phys. Rev. D 63 (2001) 114014.
9. HERMES Collaboration (F. Ellinghaus for the collaboration), hep-ex/0207029.

SPIN AND AZIMUTHAL ASYMMETRIES IN DIS

H.AVAKIAN
Jefferson Lab
12000 Jefferson Ave., Newport News, VA

1. Introduction

Single-spin asymmetries (SSA) in hadronic reactions have been among the most difficult phenomena to understand from first principles in QCD. Large SSAs have been observed in hadronic reactions for decades [1, 2]. In general, such single-spin asymmetries require a correlation of a particle spin direction and the orientation of the production or scattering plane. In hadronic processes, such correlations can provide a window to the physics of initial and final state interactions at the parton level. Recently, significant SSAs were reported in semi-inclusive DIS (SIDIS) by the HERMES collaboration at HERA [3, 4] for a longitudinally polarized target, the SMC collaboration at CERN for a transversely polarized target [5], and the CLAS collaboration at the Thomas Jefferson National Accelerator Facility (JLab)[6] with a polarized beam.

Single-spin asymmetries in SIDIS give access to distribution and fragmentation functions, that cannot easily be accessed in other ways. The list of novel physics observables accessible in SSAs includes the chiral-odd distribution functions, such as the transversity (δq) [7, 8], the *time-reversal odd* fragmentation functions, in particular the Collins function ($H_1^{\perp}$) [9], and the recently introduced [10, 11, 12, 13, 14] *time-reversal odd* distribution functions ($f_{1T}^{\perp}, h_1^{\perp}$). These latter functions arise from interference between amplitudes with left- and right-handed polarization states, and only exist because of chiral symmetry breaking in QCD. Their study, therefore, provides a new avenue for probing the chiral nature of the partonic structure of hadrons.

It is also argued that in both semi-inclusive [15] and hard exclusive [16, 17] pion production, scaling sets in for cross section ratios and spin asymmetries at lower Q^2 than it does for the absolute cross section. There are quite a few examples of remarkable agreement between spin asymmetries measured at different beam energies over a wide Q^2 range. One par-

E. Steffens and R. Shanidze (eds.), Spin Structure of the Nucleon, 95–108.

distribution functions		chirality even	chirality odd
twist 2	U	$\mathbf{q}$	$h_1^\perp$
	L	$\mathbf{\Delta q}$	$h_{1L}^\perp$
	T	$f_{1T}^\perp$ g_{1T}	$\delta\mathbf{q}$ $h_{1T}^\perp$
twist 3	U	$f^\perp$	$\mathbf{e}$
	L	$g_L^\perp$	$\mathbf{h}_L$
	T	$\mathbf{g}_T$ $g_T^\perp$	h_T $h_T^\perp$

TABLE 1. List of twist-2 and twist-3 distribution functions accessible in SIDIS.

ticular case is the very good agreement of the HERMES data with the SMC data, taken at 6-12 times higher average Q^2, showing that the semi-inclusive spin asymmetries are Q^2 independent within the present accuracy of the experiments [18]. This makes it possible for the measurement of spin-asymmetries to be a major tool for the study of different parton distribution function (PDF) measurements in the Q^2 domain of a few GeV^2.

At leading-twist, the quark structure of hadrons is described by three distribution functions: the number density $q(x)$; the helicity distribution $\Delta q(x)$, and the transversity distribution $\delta q(x)$. If the transverse momentum k_T of partons also included, the number of independent distribution functions at leading twist increases to six [19, 20] (three of which reduce to $q(x), \Delta q(x)$ and $\delta q(x)$ when integrated over k_T). Because they depend on the longitudinal and transverse momentum, these "3-dimensional" transverse momentum dependent (TMD) functions provide a more complete picture of nucleon structure. Without time invariance, two additional functions $(f_{1T}^\perp, h_1^\perp)$ are permitted, bringing the total number of distribution functions to eight. Table 1 lists the twist-2 and twist-3 distribution functions (those that survive after the k_T-integration are denoted in boldface) contributing to the double-polarized cross section in SIDIS (see [19]).

TMDs are nonperturbative functions that carry information not only on longitudinal but also on transverse hadron structure. TMD distributions in impact parameter space are correlation functions for the transverse distance of a single parton with respect to all other partons in the wave function in contrast to Generalized Parton Distributions (GPDs) for which the relative distance of partons to each other in a hadron stays the same [21].

Until recently the TMD distribution functions had mainly academic interest. They appear in azimuthal moments of double-polarized cross sections in single-hadron production in DIS [19, 20]. As shown recently in Ref.[22], the interaction of active parton in the hadron with target spectators lead to gauge-invariant TMD parton distributions. Brodsky et al. [12] discussed final state diffractive scattering, which gives rise to interference effects in the DIS cross section[23]. A non trivial phase structure of QCD amplitudes due to rescattering results in *time-reversal odd* (T-odd) effects and the appearance of single-spin asymmetries at leading twist[12, 13]. Furthermore, it was demonstrated recently that a nonzero orbital angular momentum of partons in the nucleon is crucial in forming the target SSA. The interference of different amplitudes arising from the target hadron's wavefunction which gives single-spin asymmetries [12, 24], also yields the Pauli form factor $F_2(t)$ and the GPD $E(x, \xi, t)$ entering Deeply Virtual Compton Scattering[25, 26]. The helicity-flip GPD $E(x, \xi, t)$ in impact parameter space describes how the distribution of partons in the transverse plane depends on the polarization of the nucleon and is directly linked to various transverse single spin asymmetries [27].

Both CLAS and HERMES detectors are able to access a large variety of higher-twist effects. An important feature of higher-twist distribution and fragmentation functions is that, while being important for understanding the long-range quark-gluon dynamics, they contribute at leading order in $1/Q$ to certain asymmetries[8, 19, 20, 28]. Higher-twist structure functions are important at CEBAF and HERMES energies because of the phenomenon of parton-hadron duality [29], or 'precocious scaling' [30, 31].

A key goal is to carefully study the transition between the nonperturbative and perturbative regimes of QCD using simultaneous measurements of the Q^2 and x dependencies of cross sections and beam/target spin asymmetries for different final state hadrons with extraction of the corresponding structure functions and separation of the contributions of different distribution and fragmentation functions.

2. Polarized SIDIS cross section

The total cross section for single pion production by longitudinally polarized leptons that scatter off unpolarized protons is defined by a set of structure functions. The beam spin-independent part of the cross section (σ_{UU} in Ref. [19]) arises from the symmetric part of the hadronic tensor, and the helicity-dependent part (σ_{LU})[28] arises from the anti-symmetric part:

$$\frac{d\sigma_{UU}}{dx\, dy\, dz\, d^2 P_\perp} = \frac{4\pi\, \alpha^2\, s}{Q^4} x \left\{ \left(1 - y + \frac{1}{2} y^2 + \frac{1}{4} \gamma^2 \right) \mathcal{H}_T + \left(1 - y - \frac{1}{4} \gamma^2 \right) \mathcal{H}_L \right.$$

$$\left. - (2 - y) \sqrt{1 - y - \frac{1}{4} \gamma^2}\, \cos\phi\, \mathcal{H}_{LT} \right.$$

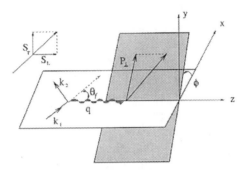

Figure 1.   Kinematics for the pion electroproduction.

$$+ \left(1 - y - \frac{1}{4}\gamma^2\right) \cos 2\phi \; \mathcal{H}_{TT} \bigg\},$$

$$\frac{d\sigma_{LU}}{dx\,dy\,dz\,d^2P_\perp} = \lambda_e \frac{4\pi\,\alpha^2\,s}{Q^4}\, x \sqrt{y^2 + \gamma^2} \sqrt{1 - y - \frac{1}{4}\gamma^2} \, \sin\phi \; \mathcal{H}'_{LT}, \qquad (1)$$

where ϕ is the azimuthal angle between the scattering plane formed by the initial (k_1) and final (k_2) momenta of the electron and the production plane formed by the transverse momentum of the observed hadron ($P_\perp$) and the virtual photon (see Fig. 1), λ_e is the electron helicity. The target mass corrections are explicitly included in the kinematics via the term $\gamma^2 = 4M^2x^2/Q^2$.

The relevant kinematical variables are defined as: $x = Q^2/2P_1 \cdot q, y = P_1 q/P_1 \cdot k_1$, $z = P_1 \cdot P/P_1 \cdot q$, where $Q^2 = -q^2$, $q = k_1 - k_2$ is the momentum of the virtual photon, and P_1 and P are the target and observed final-state hadron 4-momentum, respectively. The structure functions $\mathcal{H}_T, \mathcal{H}_L, \mathcal{H}_{TT}, \mathcal{H}_{LT}$, and $\mathcal{H}_{LT'}$ are related to the transverse and longitudinal photon contributions and their interference. The subscripts "U, L, T" stand for the unpolarized (U), longitudinally polarized (L), and transversely polarized (T) states of the beam (first index) and target (second index). Measurements of average moments $\langle W(\phi) \rangle_{BT} = \int \sigma_{BT}(\phi) W(\phi)d\phi / \int \sigma(\phi)d\phi$ ($W(\phi) = \sin\phi, \cos\phi, \sin 2\phi$) of the cross section $\sigma_{BT}^{W(\phi)}$ will single out corresponding terms in the cross section. For spin dependent moments the measurement of average moment is equivalent to the measurement of corresponding spin asymmetries A_{BT}^W Thus the SSA in the $\sin\phi$ moment of the cross section for longitudinally polarized beam and unpolarized target is defined as:

$$\frac{1}{2} A_{LU}^{\sin\phi} = \langle \sin\phi \rangle_{LU} = \frac{1}{P^\pm N^\pm} \sum_{i=1}^{N^\pm} \sin\phi_i, \qquad (2)$$

where $P^\pm$ and $N^\pm$ are polarization values and number of events for $\pm$ helicity state. The final asymmetry is defined by the weighted average over two independent measurements for both helicity states.

Additional single and double spin-dependent contributions with corresponding structure functions appear in the SIDIS cross section for polarized targets or for polarized final states. Assuming that the quark scattering process and the fragmentation process factorize, and that the fragmentation functions depend only on the fractional energy, z, the structure functions could be presented as a convolution of a distribution function and a fragmentation function. Both assumptions have yet to be experimentally confirmed at JLab and HERMES energies for different hard processes.

3. Azimuthal asymmetries with unpolarized target

Azimuthal asymmetries in SIDIS has been recognized as an important testing ground for QCD. Georgi and Politzer [32] found a negative contribution to the $\cos\phi$ moment of the cross section at first order in α_S perturbative theory. It appeared that perturbative QCD alone does not describe the observed azimuthal angular dependencies and the major part is coming from the lowest order processes due to the intrinsic transverse momentum of partons bound inside the proton [33]. The importance of pion bound-state effects in azimuthal distributions in single pion electroproduction was pointed out by Berger [34]. It is an intriguing feature of Berger's higher twist mechanism that it generates azimuthal moments opposite in sign to those of [32, 33]. In semi-exclusive electroproduction, the final meson production via hard-gluon exchange [36, 37, 38] is expected to dominate the cross section in the kinematic regime where the ejected meson picks up most of the virtual photon momentum (which implies that z is large)[37]. It is essential that in the theory of semi-exclusive reactions, the formation of final hadronic states is described in terms of quark distribution amplitudes, therefore providing a connection between inclusive and exclusive reactions (see Fig. 2).

It was shown in Ref.[35] that higher twist effects may be isolated in semi-exclusive pion production for moderate values of Q^2. Significant $\cos\phi$ and $\cos2\phi$ moments (see Fig.3) were predicted in the exclusive limit ($z \to 1$). Another important feature of semi-exclusive pion production is the suppression of π^0 with respect to π^+ in direct production [39].

It was also noted that with JLab upgraded to 12 GeV, the semi-exclusive channel can reach high virtuality of the exchanged gluon, corresponding to

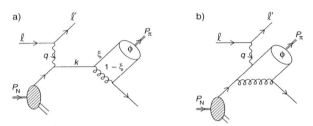

Figure 2. Leading contributions to the amplitude of the reaction $u + e^- \to e^- + \pi^+ + d$.

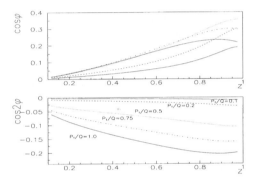

Figure 3. Azimuthal moments in unpolarized cross section.

about $Q^2 \sim 30$ GeV^2 for the exclusive case for the pion form factor [40]. An important physics implication of the connection between inclusive and exclusive reactions is that the corresponding subprocess, $\gamma^* q \to \pi q$, is an essential component of the formalism of Deeply Virtual Meson production. Semi–inclusive measurements may therefore produce model–independent information necessary to extract (polarized) Generalized Parton Distributions from deeply–virtual exclusive electroproduction of mesons.

Both CLAS and HERMES detectors can provide complete kinematic coverage of semi-exclusive electroproduction reactions in the deep-inelastic region.

4. Unpolarized Target, Polarized Beam

Assuming factorization of the quark scattering and fragmentation processes, the distribution and fragmentation functions responsible for a non-zero $\mathcal{H}'_{LT}$ in SIDIS were first identified by Levelt and Mulders [28]. They include the twist-3 unpolarized distribution function $e(x)$ introduced by

Jaffe and Ji [8], and the polarized fragmentation function $H_1^\perp(z)$ first discussed by Collins [9]. The x dependence of beam SSA is defined by the ratio of the twist-3 unpolarized distribution function $e(x)$ and the leading twist distribution function $f_1(x)$ [28]. The expression for the beam SSA involves the convolution of the Collins function and the interaction dependent part $\tilde{e}(x)$ of the higher twist function $e(x)$:

$$\sigma_{LU}^{\sin\phi} \propto \lambda_e 2y\sqrt{1-y}\sin(\phi)\sum_{q,\bar{q}} e_q^2 x^2 \tilde{e}(x) H_1^{\perp q}(z), \qquad (3)$$

where the sum $\sum_{q,\bar{q}}$ is over quark flavors, $e^q(x) = \tilde{e}^q(x) + m_q/M\, f_1^q(x)/x$ and m_q is the current quark mass.

First measurement of significant beam-spin asymmetries was performed using a 4.25 GeV longitudinally polarized electron beam and the CEBAF Large Acceptance Spectrometer (CLAS) [41] at Jlab. The data analysis reveals very significant single-beam spin asymmetries in pion azimuthal distributions in deep-inelastic electroproduction ($Q^2 > 1$ GeV^2,$W^2 > 4$ GeV^2) at large z [6]. The $A_{LU}^{\sin\phi}$ averaged over two spin states as a function of z is plotted in Fig. 4. Observed beam SSA are increasing with z. The dashed line (see Fig. 4) shows the z-dependence of the target-SSA measured by the HERMES collaboration [3]. The preliminary analysis of target SSA(A_{UL}) with the CLAS polarized target (see Fig. 9) is in agreement with the HERMES result, confirming the striking difference in the z-dependence of beam and target SSA.

The measured beam SSA $A_{LU}^{\sin\phi}$ is positive for a positive electron helicity in the range of $0.15 < x < 0.4$. The relatively flat x dependence of the measured SSA for $0.15 < x < 0.4$ (see Fig. 4) is in qualitative agreement with an existing calculation of $e(x)$ in the Bag model at Q^2=1 GeV^2[42]. The first extraction of the twist-3 distribution function from CLAS data [43] is shown in Fig. 5. With a certain approximation for the twist-3 function $e(x)$, the beam SSA could become a major source of information on the T-odd polarized fragmentation function.

Unlike target SSAs (Fig. 9) the beam SSA does not change sign (Fig. 4) at large z. The z dependence of beam SSA for exclusive events in the DIS region is shown on the Fig. 6. It is positive at large z and compatible with the A_{LU} for the semi-exclusive sample at the same value of z. The A_{LU} for exclusive events changes sign at low z where the the majority of pions are not expected to contain the struck quark.

The SSAs measured in DIS are related to ratios of corresponding structure functions. However the single-spin observables $A_{LU}, A_{UL}, \ldots$ contain also kinematic factor that depends on y, the fraction of energy of initial

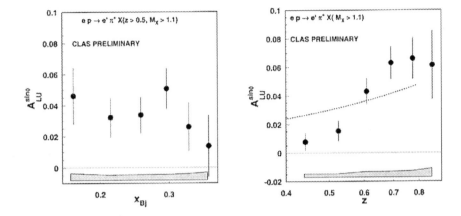

Figure 4.   The beam-spin azimuthal asymmetry ($\sin\phi$ moment of the cross section) extracted from hydrogen data at 4.25GeV as a function of x in a range $0.5 < z < 0.8$ (left panel) and z in a range $0.15 < x < 0.4$ (right panel). The bars show the statistical uncertainty and the band represents the systematic uncertainties.

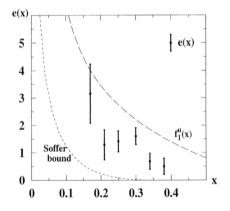

Figure 5.   The flavor combination $e(x) = (e^u + \frac{1}{4}e^{\bar{d}})(x)$ vs. x, extracted from the CLAS beam-spin azimuthal asymmetry. The error bars are due to statistical errors of the CLAS data with $\langle Q^2 \rangle = 1.5\,\text{GeV}^2$. A fit to published HERMES data[3] on target SSA was used in the parameterization of the Collins function. For comparison f_1^u and the twist-3 Soffer bound are shown.

lepton carried by the virtual photon (see Eq. 1). To extract the ratio of distribution and fragmentation functions one has to remove the kinematic factor. The kinematic factor in A_{LU} is shown in Fig. 7 as a function of x and z for HERMES and CLAS. While it is constant for the CLAS data sample,

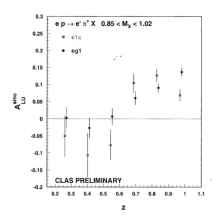

Figure 6.   The $\sin\phi$ moment ($A_{LU}^{\sin\phi}$) for exclusive π^+ from NH$_3$ (eg1) and hydrogen (e1c) runs. Error bars show only the statistical uncertainty.

it has some dependence on x and z for HERMES due to the lower cut on the pion energy used in the pion identification at HERMES in 1996/97.

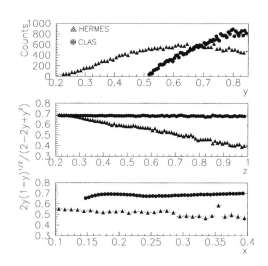

Figure 7.   The y-distributions at CLAS and HERMES (upper plot) and the dependence of the kinematic factor on x and z.

Beam SSA measurements do not require polarized targets and are free of dilution. They could be measured at the highest accessible luminosities.

This makes them an important tool for factorization studies using measurements of different final state hadrons. In particular measurement of SSAs with K^+ in comparison with π^+ would provide an experimental test of factorization and u-quark dominance. In addition in this ratios one tests the *universality* property of factorization: since the PDFs are the same in both processes, the x-dependence of the asymmetries for π^+ and K^+ must be identical if the factorization holds [44].

5. Polarized Targets

For polarized targets, several azimuthal asymmetries already arise at leading order. The following contributions were investigated in Refs. [9, 19, 20, 12, 14, 45]:

$$\sigma_{LT}^{\cos\phi} \propto \lambda_e S_T y(1-y/2)\cos(\phi-\phi_S)\sum_{q,\bar{q}} e_q^2 x g_{1T}^q(x) D_1^q(z), \qquad (4)$$

$$\sigma_{UL}^{\sin 2\phi} \propto S_L 2(1-y)\sin 2\phi \sum_{q,\bar{q}} e_q^2 x h_{1L}^{\perp q}(x) H_1^{\perp q}(z), \qquad (5)$$

$$\sigma_{UT}^{\sin\phi} \propto S_T(1-y)\sin(\phi-\phi_S)\sum_{q,\bar{q}} e_q^2 x \delta q(x) H_1^{\perp q}(z),$$

$$+ \quad S_T(1-y)\sin(\phi+\phi_S)\sum_{q,\bar{q}} e_q^2 x f_{1T}^{\perp q}(x) D_1^q(z), \qquad (6)$$

where ϕ_S is the azimuthal angle of the transverse spin in the photon frame and $D_1^q(z)$ is the spin-independent fragmentation function.

The latter two equations above describe single-spin asymmetries involving the first moment of the Collins fragmentation function integrated over the transverse momentum of the final hadron. The leading-twist SSA $\sigma_{UL}^{\sin 2\phi}$ is kinematically suppressed at low x compared to the sub-leading $\sin\phi$ moment [46]. A recent measurement of the σ_{UL} contribution by HERMES[3] is consistent with a zero $\sin 2\phi$ moment. However, at the large x values accessible at JLab, the $A_{UL}^{\sin 2\phi}$ asymmetry is predicted [46] to be large (see Fig. 8). The leading-twist distribution function $h_{1L}^{\perp}(x)$, accessible in that measurement, describes the transverse polarization of quarks in a longitudinally polarized proton.

The $\sin\phi$ moment of the SIDIS cross section with a transversely polarized target (σ_{UT})[48] contains contributions both from the Sivers effect (T-odd distribution)[10] and the Collins effect (T-odd fragmentation)[9]. Contributions to transverse SSAs from T-odd distributions of initial quarks ($f_{1T}^{\perp q}(x)$ term) and T-odd fragmentation of final quarks ($H_1^{\perp q}(z)$ term) could be separated by their different azimuthal and z-dependencies. The

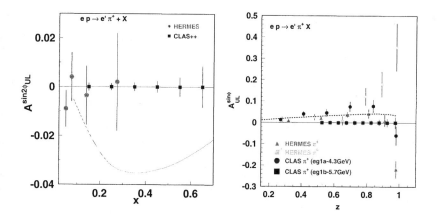

Figure 8. Dependence on x of the longitudinally polarized target SSA, $A_{UL}^{\sin 2\phi}$. Circles are HERMES data for $A_{UL}^{\sin 2\phi}$, and squares represent expected statistical errors from CLAS at 11 GeV with 2000 hours of data taking. The curve is the prediction from Ref. [46].

Figure 9. $A_{UL}^{\sin\phi}$ as a function of z from CLAS compared to HERMES [3, 4]. Squares are projection for polarized NH_3 target running of CLAS at 5.7 GeV (EG1 experiment). The curve is the prediction from [47] for HERMES kinematic conditions.

azimuthal angles entering the expression for both contributions are shown in Fig. 10. In the case of a longitudinally polarized target the Collins (ϕ_C) and Sivers (ϕ_B) angles have opposite sign. Though one can study both contributions measuring the θ_γ dependence of target SSA for longitudinal target polarization, the definitive resolution could be achieved only with transversely polarized target measurements.

Assuming that the transversity of the sea is negligible ($\delta\bar{q} = 0$) and ignoring the non-valence quark contributions in positive pion production, the single-spin transverse asymmetry arising from fragmentation becomes:

$$A_{UT}^{\pi^+} \propto \frac{4\delta u(x)}{4u(x) + \bar{d}(x)} \frac{H_1^{\perp u \to \pi^+}(z, P_\perp)}{D_1^{u \to \pi^+}(z, P_\perp)}, \qquad (7)$$

The target single-spin asymmetry from polarized quark fragmentation extracted for CLAS kinematics at 6 GeV is plotted in Fig. 11. The estimate was done assuming $\delta q \approx \Delta q$ and an approximation for the Collins fragmentation function from Ref.[46]. Additional cuts were applied on z ($z > 0.5$) and the missing mass of the $e'\pi^+$ system ($M_X(\pi^+) > 1.3$ GeV). The curves have been calculated assuming a luminosity of $10^{34} \mathrm{cm}^{-2}\mathrm{s}^{-1}$, with a NH_3 target polarization of 85% and a dilution factor 0.176, with 2000 hours of data taking. The asymmetry is integrated over all hadron transverse mo-

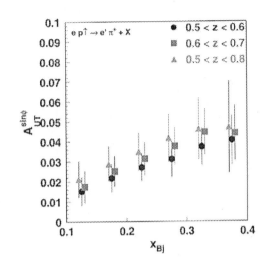

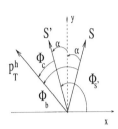

Figure 10. Defini-
tion of Collins (ϕ_C)
and Sivers (ϕ_B) an-
gles. The S and S'
are initial and fi-
nal quark polariza-
tions in transverse to
virtual photon direc-
tion plane.

Figure 11. Projected transverse spin asymmetry ($A_{UT}^{\sin\phi}$) in
single π^+ production with CLAS at 5.7 GeV.

menta. The extraction of the transversity from $A_{UT}^{\sin\phi}$ could be performed
via Eq. 7 using parameterizations for the unpolarized distribution functions
$u(x)$ and $\bar{d}(x)$.

The SIDIS cross section with a longitudinally polarized target in the
sub-leading order contains an additional contribution to the $\sin\phi$ moment
(σ_{UL}) [45, 48, 47]:

$$\sigma_{UL}^{\sin\phi} \quad \propto \quad S_L \sin\phi \, (2-y)\sqrt{1-y}\frac{M}{Q}\sum_{q,\bar{q}} e_q^2 x^2 h_L^q(x) H_1^{\perp q}(z). \qquad (8)$$

The $\sin\phi$ moment of the cross section measured at CLAS is in good agreement with the HERMES measurement, which indicates that the asymmetry
observables are not sensitive to the beam energy (see Fig.9). More data are
needed to separate different contributions [17, 12, 49, 50] resulting in a
sign flip of the target SSA at large z observed by HERMES[3]. The data
available from polarized NH_3 running with 5.7 GeV at CLAS (see Fig.9)
would allow the extraction of target-SSA as a function of the angle of the
virtual photon with the beam direction and separation of transverse and
longitudinal spin contributions.

The target SSA measured by HERMES [3, 4] and analyzed in terms
of the fragmentation effect, in addition to a contribution from the Collins
function, contains other contributions which in a certain kinematic range

might be significant [49, 50]. When neglecting the interaction dependent part in distribution function h_L,

$$A_{UL}^{\sin\phi} \propto x\, \frac{h_L}{f_1}[H_1^{\perp(1)}(z) + \frac{\tilde{H}(z)}{2z}]/D(z).$$

The HT term with $\tilde{H}(z)$ gives a contribution with opposite sign [50] and the contamination from HT fragmentation function at large z depends significantly on the shape of the Collins function [49]. Recently, the Collins function for pions was calculated in a chiral invariant approach at a low scale [15] and it was shown that at large z the function rises much faster than previously predicted [46, 47] in the analysis using the HERMES data on target SSA. It was also pointed out that the ratio of polarized and unpolarized fragmentation is almost scale independent[15]. Another important source of possible contribution to SSA with a longitudinally polarized target at large z are exclusive transverse SSA asymmetries [16, 17] also predicted to be large enough to contribute through the small angle of the virtual photon with the proton spin direction (see Fig.1). All this makes the beam SSA a cleaner observable for extraction of the Collins fragmentation function at large z, where the analyzing power is large.

Single-spin asymmetry measurements open up a unique possibility to access T-odd distribution functions in semi-inclusive DIS [12, 13]. Ongoing experiments with 6 GeV electrons at CLAS, 27.5 GeV positrons at HERMES and 160 GeV muons at COMPASS, using various polarized and unpolarized targets will study the flavor dependence of the corresponding functions[51, 52]. Significantly higher statistics will enable the extraction of the x and Q^2 dependencies for different azimuthal moments in a wide kinematical range allowing to reveal the source of the observed asymmetries and extract underlying distribution functions.

References

1. K. Heller *et al.* 'Proceedings of Spin 96',Amsterdam,Sep.1996,p23
2. Fermilab E704 collaboration (A. Bravar *et al.*), Phys.Rev.Lett. **77**, 2626 (1996).
3. HERMES collaboration (A. Airapetyan *et al.*), Phys.Rev.Lett. **84**, 4047 (2000).
4. HERMES collaboration (A. Airapetyan *et al.*), Phys.Rev. **D64**, 097101 (2001).
5. A. Bravar, Nucl. Phys. (Proc. Suppl.) **B79** 521 (1999).
6. CLAS Collaboration (H. Avakian *et al.*) in preparation.
7. J. Ralston and D. Soper, Nucl. Phys. **B152**, 109 (1979)
8. R.L. Jaffe and X. Ji Nucl.Phys. **B375** (1992) 527.
9. J. Collins, Nucl. Phys. **B396**, 161 (1993).
10. D. Sivers, Phys.Rev. **D43**, 261 (1991).
11. M. Anselmino and F. Murgia, Phys. Lett. B **442**, 470 (1998).
12. S. Brodsky *et al.* Phys. Lett. B **530**, 99 (2002). hep-ph/0201165.
13. J. Collins, hep-ph/0204004.
14. X. Ji, F. Yuan e-Print Archive: hep-ph/0206057.

15. A. Bacchetta *et al.* Phys.Rev., **D65**, 94021 (2002).
16. L. Frankfurt *et al.*, Phys. Rev. D **60** 014010, (1999).
17. L. Frankfurt *et al.*,, Phys. Rev. Lett., **84** 2589 (2000).
18. HERMES Collaboration (K. Ackerstaff *et al.*) Phys.Lett. **B464**, 123 (1999).
19. P.J. Mulders and R.D. Tangerman, Nucl. Phys. **B461**, 197 (1996).
20. A. Kotzinian, Nucl. Phys. **B 441**, 234 (1995).
21. M. Diehl, hep-ph/0205208.
22. A. Belitsky,X. Ji and F. Yuan hep-ph/0208038.
23. S. Brodsky *et al.*, hep-ph/0104291
24. S. Brodsky *et al.*, hep-ph/0206259.
25. X. Ji, Phys. Rev. Lett. **78**, 610 (1997); Phys. Rev. D **55**, 7114 (1997).
26. A.V. Radyushkin, Phys. Lett. **B380**, 417 (1996); Phys. Rev. D **56**, 5524 (1997).
27. M. Burkardt, hep-ph/02091179.
28. J. Levelt and P. J. Mulders, Phys. Lett. **B338**, 357 (1994).
29. E.D. Bloom and F.J. Gilman, Phys.Rev. D **4**, 2901 (1972),
30. A. De Rujula,H. Georgi and H.D. Politzer, Ann.Phys. **103**, 315 (1977).
31. X. Ji and P. Unrau, Phys.Rev. D **52**, 72 (1995).
32. H. Georgi and H.D. Politzer, Phys.Rev.Lett., **40**, 3 (1978).
33. R. Cahn, Phys.Lett., **B78**, 269 (1978); Phys.Rev., D **40**, 3107 (1989).
34. E. Berger, Z. Phys., **C 4**, 289 (1980).
35. A. Brandenburg, V. Khoze, D. Muller Phys. Lett. B **347**, 413 (1995)
36. A. Brandenburg, V. V. Khoze and D. Muller,Phys. Lett. B **347**, 413 (1995).
37. A. Afanasev *et al.*, Phys. Lett. B **398**, 393 (1997); Phys. Rev. D **58**, 054007 (1998).
38. S. J. Brodsky, M. Diehl, P. Hoyer and S. Peigne, Phys. Lett. B **449**, 306 (1999)
39. A. Afanasev *et al.* hep-ph/9701215.
40. A. Afanasev *et al.*, Phys. Rev. D **62**, 074011 (2000).
41. B. Mecking *et al.*, "The CLAS Detector", in preparation.
42. M. Signal, Nucl. Phys. B **497** (1997) 415 [arXiv:hep-ph/9610480].
43. A. Efremov *et al.* Proceedings of QCD'N02 Ferrara 2002, hep-ph/0206267.
44. M. Diehl and R. Jakob, private communication.
45. A.M. Kotzinian and P.J. Mulders, Phys. Rev. **D54** 1229 (1996).
46. A. Efremov K. Goeke, P. Schweitzer Phys.Lett. **B552**, 37 (2001).
47. A. M. Kotzinian *et al.*, Nucl.Phys. **A666**, 290-295 (2000).
48. D. Boer and P. Mulders, Phys.Rev. **D57**, 5780 (1998).
49. P. J. Mulders, R. M. Boglione, Phys.Lett B **478**, 114 (2000).
50. H. Avakian, Proceedings of DIS-2000, Liverpool University 2000.
51. A. Efremov *et al.*, hep-ph/0001214,Czech.J.Phys Suppl.
52. Bo-Qiang Ma *et al.*, hep-ph/0110324.

NUCLEAR ATTENUATION IN SIDIS

N. BIANCHI
INFN-Laboratori Nazionali di Frascati
Via E. Fermi 40 - I00044 Frascati - Italy

1. Introduction

The study of nuclear medium effects with electromagnetic probes takes advantage from the precise knowledge of the direct coupling of the e.m. field with the charged nuclear currents which have to be investigated. In addition, the small e.m. interaction cross section allows to explore the whole volume of the nucleus, thus photo-producing fast moving quarks and the subsequent formed hadrons in high static nuclear densities.

The inclusive Deep-Inelastic-Scattering (DIS) of high energy leptons on free nucleons has been used from late 60s to measure the partonic distributions and, by using nuclear targets, to study the medium modification to these distributions. Nuclear modifications of partonic distributions were found to be large and were explained by invoking medium effects at both hadronic and partonic levels (see [1] for a review).

Semi-Inclusive DIS (SIDIS) of high energy leptons on free nucleons can be used to study another fundamental function of the quark, the *fragmentation function*. This function can be also modified by the nuclear medium and only recently this effect has been clearly revealed by new and precise data.

2. Fragmentation functions and multiplicities

The elementary SIDIS diagram is shown in Fig. 1. This process can be factorized in a parton distribution function q_f, which is function of the Bjorken scaling variable $x = Q^2/(2M\nu)$, a hard scattering cross section $d\sigma_f$, which is calculable in QCD, and a parton fragmentation function.

The fragmentation function $D_f^\pi(z)$ represents the probability for a quark of flavor f to fragment into a specific hadron (a pion in this case) with relative energy $z = E_\pi/\nu$ with respect to the initial one of the struck quark.

E. Steffens and R. Shanidze (eds.), Spin Structure of the Nucleon, 109–120.

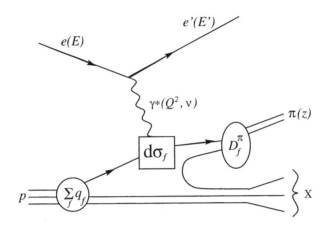

Figure 1. Diagram of Deep Inelastic Scattering on proton with the exchange of a virtual photon of 4-momentum Q^2 and energy $\nu = E - E'$. In Inclusive DIS only the scattered lepton e' is detected. In Semi-Inclusive DIS also the leading hadron (π in this case) which is formed from the struck quark is detected with a fractional energy z with respect to ν. The target fragments X are not observed in Semi-Inclusive DIS.

In analogy with the structure functions, the fragmentation functions are function of a scaling variable z and their logarithmic Q^2-evolution is determined by the DGLAP equations. Fragmentation functions are universal and do not depend from the generating process (e^+e^-, $p\bar{p}$, ep, γp, $\gamma\gamma$), apart from the appropriate choices of the different scales (renormalization, factorization and fragmentation scales) that enter in the processes.

Fragmentation functions are not calculable in QCD, but several phenomenological models are used to describe the mechanism of the hadron formation. In the Cluster model, color singlet clusters of $q\bar{q}$ are formed before decaying isotropically into hadron pairs. In the String model, a color flux between the initial $q\bar{q}$ is generated and after the string breaks up into hadrons via $q\bar{q}$ (for a review see [2]). The cleanest way to measure fragmentation functions is in the $e^+e^- \rightarrow hadrons$ reaction, but in this case one can only measure the fragmentation into hadron pairs (for example into $\pi^+\pi^-$) with no possible separation of the *favored* (like for $u \rightarrow u\bar{d} \equiv \pi^+$) and *unfavored* (like for $u \rightarrow d\bar{u} \equiv \pi^-$) ones. Fragmentation functions also obey to simple isospin invariance rules, like $D_u^{\pi^+} \equiv D_d^{\pi^-}$.

In the Quark Parton Model (QPM) description of SIDIS, the fragmentation functions are multiplied by the structure functions and experimentally one can determine hadron multiplicities by normalizing the SIDIS yield to the DIS rate:

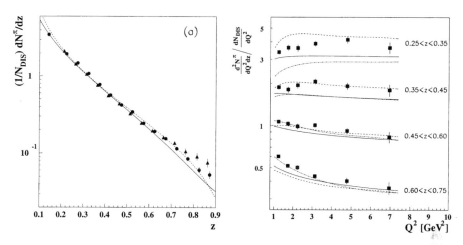

Figure 2. Left panel : SIDIS multiplicities measured by HERMES for average charged (triangles) and neutral (circles) pions. The solid curve is the Field-Feynmann parameterization for the pion fragmentation function at fixed scale. Dashed curve is a fit to the π^0 data. Right panel: Q^2-evolution of the total (charged + neutral) pion multiplicity as measured by HERMES in different bins of z. Curves are NLO-QCD evolution of fragmentation functions [3] which are fitted mainly to LEP results in hadron production. The solid curve is a fit to most recent LEP data. From [4].

$$\frac{1}{N_{DIS}}\frac{dN^h(x,z)}{dz} = \frac{\sum_f e_f^2 q_f(x) D_f^h(z)}{\sum_f e_f^2 q_f(x)}$$

It is easy to demonstrate that within the QPM and under the u-quark dominance assumption, the pion multiplicities are almost equivalent to the pion fragmentation functions.

In QCD the relation between fragmentation functions and multiplicities is more complicated since both the LO and the NLO terms of the cross section are convoluted with the initial photon and parton distributions and with the final hadron fragmentation functions [3].

Nevertheless, it has been shown that, once integrating over a broad range of x, the multiplicities are a good approximation of the fragmentation functions [4]. Fig.2 shows the comparison of multiplicities for pions, as measured in SIDIS in the HERMES experiment, with a calculation at a fixed scale (left panel) and with the NLO-QCD evolution of fragmentation functions (right panel). The latter has been determined from LEP measurements which have been performed with time-like virtual photons of about 4 orders of magnitude higher with respect to the space-like virtual photons in the HERMES experiment. The nice agreement between data and curves suggests the validity of the factorization assumption of the SIDIS process,

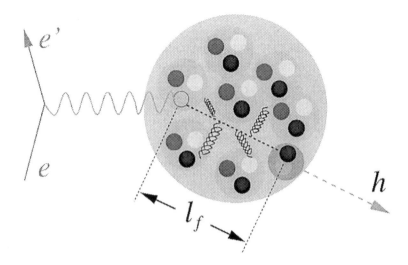

Figure 3.   Pictorial view of SIDIS in nuclei. The observed hadron yield is reduced and modified by the interaction of the struck quark and of the formed hadron inside the nucleus.

as described in Fig. 1, and the reasonable equivalence of fragmentation functions and hadron multiplicities.

3. Hadronization in the nuclear medium

The study of the medium modification of fragmentation functions can be performed in SIDIS on nuclear targets. As illustrated in Fig.3, the nucleus acts as an ensemble of targets which reduce the multiplicity of fast hadrons due to both partonic and hadronic interactions. The partonic and hadronic effects are discriminated according to the time $\tau_h = l_h/c$ of formation of the hadron which may occur inside or outside the nucleus. The hadron formation times, which are basic ingredients also in hadron-nucleus and in heavy-ion reactions, are currently not well determined. In particular, the study of the Quark Gluon Plasma requires the knowledge of the proper time at which the plasma is formed and this time is generally assumed to be comparable to the hadron formation time. Typically, for hadrons composed by light quarks, the simple expression $\tau_h \approx E R_h^2$ is used, where R_h is the hadron radius. This expression suggests a formation time of $\tau_\pi \sim$ 1-2 $(E/GeV) \cdot fm/c$ for pions.

After the first measurements performed at SLAC [5] and CERN [6], the more recent results from HERMES [7, 8] provided a new strong boost of interest in this field. HERMES is an experiment at DESY which is mainly devoted to the study of the spin structure of the nucleon by using a polarized

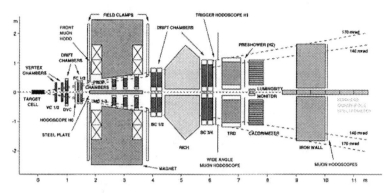

Figure 4. Side view of the HERMES spectrometer at DESY. The HERA positron (or electron) beam enters from the left and scatters into the target cell. The scattered positron is detected together with the produced leading hadron by the large angle magnetic spectrometer. Momentum is reconstructed with an accuracy of $\sim 1\%$ by 57 tracking planes before, in and after the large dipole magnet. The tracking system allows the vertex reconstruction and the angular determination with an accuracy better than 0.6 mrad. Scattered positrons are identified with the Calorimeter, the TRD and the preshower detectors. The charged hadron type identification is performed with the dual radiator RICH detector, which was installed in 1998 replacing the previous threshold Cerenkov gas detector. Neutral particles are identified with the electromagnetic calorimeter.

lepton beam and polarized internal $\vec{H}$, $\vec{D}$ and $^3\vec{He}$ gas targets.

However, some HERMES runs have been performed with nuclear gas targets $(D, {}^4He, N, Ne, Kr)$ with densities up to $\sim 10^{16}$ $nucl \cdot cm^{-2}$, allowing the extraction of interesting informations on the hadron multiplicities in the nuclei. The layout of the HERMES spectrometer [9] is shown in Fig.4. With an incident positron (or electron) beam of 27.5 GeV and with the large acceptance of the spectrometer, which covers the polar angle $0.04 \div 0.22$ rad, the kinematic region explored by HERMES is about $0.004 < x < 0.8$ and $0.2 < Q^2 < 10$ GeV^2. This region is slightly reduced by additional constrains ($W > 2$ GeV and $Q^2 > 1$ GeV^2) which are needed to ensure the DIS regime. The large acceptance and the complete particle identification make HERMES well suited for several semi-inclusive reactions. To minimize the hadron produced in the target fragmentation and to ensure the factorization of the SIDIS process, only fast hadrons with $z > 0.2$ are considered.

In the left panel of Fig.5, the HERMES results for the fast hadron multiplicity ratios of N and Kr targets with respect to the deuterium one are shown in comparison with data from earlier experiments. As it is seen, the HERMES kinematic is ideal to study these effects and the results indicate an increase of the multiplicity ratio (thus a decrease of the medium effect) with ν, in agreement with the EMC data taken at higher energies. The

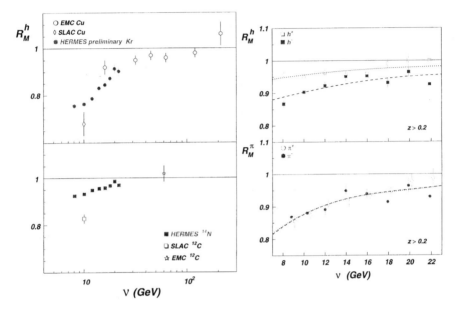

Figure 5. Left panel : multiplicity ratios as a function of ν for hadrons with $z > 0.2$ for HERMES and EMC. SLAC points are to SIDIS ratios. Right panel : HERMES multiplicity ratio $^{14}N/D$ for charged hadrons including pion (top panel) and for identified charged pion (bottom panel) as a function of ν. Curves are simple phenomenological fits to the data.

apparent discrepancy with SLAC is partially due to the EMC-effect, not considered in the SLAC data which were normalized to the luminosities instead of the DIS cross sections as in the multiplicity ratios.

In the right panel of Fig.5, the HERMES results on ^{14}N for hadrons and for pions of opposite charges are presented. While the results for π^+ and π^- are similar, a significant difference was observed between positive and negative hadrons. This has been interpreted by different attenuation of other hadrons such as protons, antiprotons and kaons which populate in different ways the positive and negative hadron samples.

The complete particle identification in HERMES, with the installation of the RICH detector in 1998, allowed to fully disentangle the information for different hadron types. This is shown in Fig.6 where the multiplicity ratios for the ^{84}Kr target are presented as function of the virtual photon energy ν (left panel) and of the scaling variable z (right panel). As it is seen, the medium effects for π^+ and π^- are equal. Also medium effects on K^- production are similar to the charged pion case. Quite interesting and unexpected is the difference between K^+ and K^- and especially between p and $\bar{p}$. In particular the effect for the proton is strongly different at low-z, where a contamination from the target fragmentation is more probable.

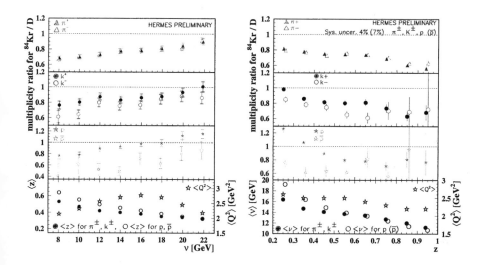

Figure 6. HERMES preliminary multiplicity ratio $^{84}Kr/D$ for different charged hadrons as a function of the virtual photon energy ν (left panel) and of the scaling variable z(right panel).. In the lower panels the relevant average kinematic variables are presented for each ν- or z-bin.

The different results for different hadrons may reveal differences in the modification of q and $\bar{q}$ fragmentation functions, in the hadron formation times and in the different hadronic interaction cross sections.

4. Modification of fragmentation functions

The new HERMES results on nuclear multiplicity ratios have been interpreted in terms of modification of fragmentation functions.

In Ref. [10] the modification of quark fragmentation functions and their QCD evolution are described in the framework of multiple parton scattering and induced gluon radiation with additional higher-twist terms in the DGLAP equations. The basic diagrams for the propagation of the struck quark in the nuclear medium are shown in the left panel of Fig.7. Diagram (a) represents the re-scattering with gluons without gluon radiation. This diagram is responsible for the p_t broadening of the leading hadron, linearly with the nuclear dimension ($\sim A^{1/3}$). Diagram (b) represents the re-scattering with quarks without gluon radiation. This diagram is also important since mix the q, $\bar{q}$ and gluon fragmentation. The gluon radiation diagram (c) provides the dominant higher twist contribution to the QCD evolution. The coherent propagation of the emitted gluon and the lead-

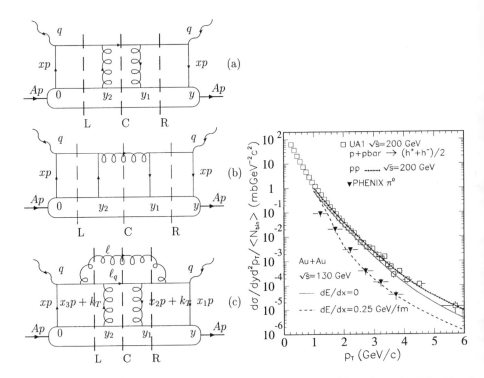

Figure 7. Left panel : diagrams for re-scattering with gluons (a) and quarks (b) without and with gluon radiation (c) in SIDIS on nuclei. Possible cuts are shown by the dashed lines (from [10]). Right panel : pQCD parton model calculation of the charged hadron and pion spectra in central $Au + Au$ collisions compared with the effective modified fragmentation function with an energy loss of $dE/dx \approx 0.25$ GeV/fm (from [11])

ing quark through the nucleus induces large interference effects which are proportional to the square of the nuclear dimension ($\sim A^{2/3}$).

As consequence of the partonic energy loss Δz, due to the gluon radiation, the fragmentation function is modified as

$$\tilde{D}(z) \approx \frac{1}{1 - \Delta z} \cdot D\left(\frac{z}{1 - \Delta z}\right)$$

In Fig. 8 these higher twist predictions are compared with HERMES data. The model has only one free parameter which correspond to the quark-gluon correlation strength inside nuclei. In the framework of this calculation, for a cold and static system like the nucleus, the quark energy loss involved in SIDIS reaction has been determined and resulted in $dE/dx \approx 0.3$ GeV/fm for the Kr nucleus. This result has been compared with the energy loss ($dE/dx \approx 0.25$ GeV/fm, see the right panel of Fig.7)

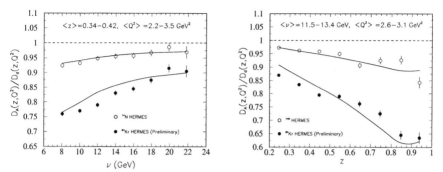

Figure 8. Predicted nuclear modification of quark fragmentation functions compared with HERMES data (from [11])

in a hot, dense and but expanding system like the one involved in heavy ion reaction at RHIC energies. An increase of the gluon density of a factor ~ 15 has been derived for the initial hot stage of $Au + Au$ reaction with respect to the one in a static nucleus [11].

In Ref. [12] the modification of quark fragmentation functions have been considered according to the Q^2-rescaling model, originally developed to interpret the EMC-effect in the nuclear structure functions. In this work the multiplicity in the medium included a rescaling of both structure and fragmentation functions

$$q_f^A(x, Q) = q_f(x, \xi_A(Q)Q)$$

$$D_f^{h|A}(z, Q) = D_f^h(x, \xi_A(Q)Q) \ ,$$

where the deconfinement scale λ_A is assumed to be proportional to the degree of overlap of the nucleons inside the given nucleus, and

$$\xi_A(Q) = (\lambda_A/\lambda_0)^{\frac{1}{2}\frac{\alpha_s(Q_0)}{\alpha_s(Q)}}$$

In Fig. 9 the rescaling predictions are shown to be in good agreement with EMC data for Cu and with the HERMES results for N. In the Kr case for the HERMES kinematic, the contribution of the hadron re-scattering inside the nucleus in clearly needed to reproduce the data.

In Ref. [13], the combined effect of the fragmentation function modification due to the nuclear induced gluon radiation and of the hadron interaction inside the nucleus has been calculated. In this work, a pion formation

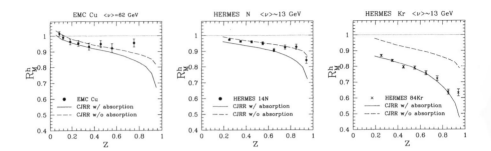

Figure 9. Comparison of EMC and HERMES data with absorption effects with rescaling only (dashed lines) and with rescaling and absorption (solid lines), (from [12]).

time of $\sim 1.2\ (E/GeV) \cdot fm/c$ has been determined in the comparison with HERMES data. This promising model is only valid for very fast pions ($z > 0,5$) and further investigations are needed for the studies of the formation of heavier mesons and baryons.

Other more phenomenological works [14] based on string models, have been developed in order to reproduce the experimental data by varying the fit parameters like the string constant and the hadronic cross sections.

5. Modification of p_t distribution

As mentioned before, the higher-twist diagrams related to the scattering of quark and gluon without gluon radiation increase the transverse momentum k_t of the struck quark with respect to its intrinsic value and, as consequence, the transverse momentum p_t of the leading hadron. This effect is shown in the left panel of Fig.10. The HERMES data for $p_t^2 < 1$ GeV^2 indicate the hadron attenuation due to the fragmentation function modification while the data for $p_t^2 \geq 1$ GeV^2 suggest the onset of a hard component due to incoherent parton scattering. The increase of p_t is larger for the heavier nucleus.

The p_t broadening is known as *Cronin effect* [15, 16] and a similar enhancement of hadron yield in nuclei at large p_t has been observed in many hadronic reactions. In pA collisions the incoming proton gains extra transverse momentum due to random soft collisions and the partons enter the final hard process with extra k_t proportional to the number of effective nucleon-nucleon collisions. Also in high energy heavy-ion collisions performed at CERN SPS, a similar effect has been clearly observed (see right panel of Fig.10) and has been described by multiple parton scattering within the Glauber formalism. It is worth to note that in the heavy-ion in the SPS regime the parton re-scattering still dominates the attenuation at

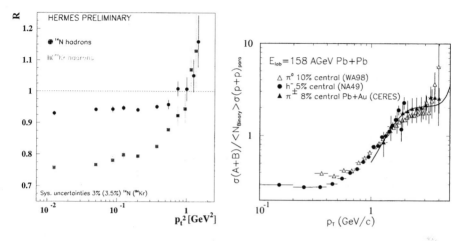

Figure 10. p_t-dependence of nuclear multiplicity ratio in SIDIS (left) and of nuclear modification factor in heavy ion reactions (right). The curve is a microscopic calculation which considers multiple parton scattering

large p_t, while the more recent RHIC data showed a clear *jet quenching* in the same large p_t-region due to the much higher density reached in the collision.

If the p_t broadening in SIDIS is mainly dependent on the multi-parton correlation function, it should be proportional to the nuclear size. Therefore it will be important to measure it on several nuclei and for different hadron types. Its direct measurement will also provide a good test of the factorization ansatz.

6. Conclusions

After the first studies of SLAC and EMC, the recent and precise HERMES results in SIDIS on N and Kr nuclei have shown a significant nuclear attenuation for fast hadrons in a new region of the kinematic plane. In addition, for the first time, different medium effects have been observed for different hadron types. In particular while the effect for the pion do not depend on the charge, a sizable difference has been observed between K^+ and K^- and between p and $\bar{p}$. The so called Cronin effect in the p_t broadening of the leading hadron has also been observed in SIDIS.

These results have been interpreted in terms of medium modification of fragmentation functions due to gluon bremsstrahlung and re-scattering of the struck quark while traveling in the medium. However the data on Kr suggested the importance of the additional hadron re-interaction in the nucleus.

The SIDIS results provide important informations on the parton energy loss and on the hadron formation time which are basic ingredients for other hadronic reactions and in particular for the heavy-ion interactions at high energies, which are currently studied to provide hints for the Quark Gluon Plasma formation.

References

1. Arneodo M. (1994) *Phys. Rept.* **240** 301.
2. Webber B.R. (2000) *Int.J. Mod. Phys.* **A 15S1** 577.
3. Kniehl B., Kramer G. and Potter B. (2000) *Nucl. Phys.* **B 582** 514. Binnewies J., Kniehl B. and Kramer G. (1995) *Z. Phys.* **C 65** 471 and *Phys. Rev.* **D 52** 4947.
4. HERMES Collaboration : Airapetian A. (2001) *Eur. Phys. J.* **C 21** 599.
5. Osborne L. *et al.* (1978) *Phys. Rev. Lett.* **40** 1624.
6. EMC Collaboration : Ashman J. *et al.* (1991) *Z. Phys.* **C 52** 1.
7. HERMES Collaboration : Airapetian A. (2001) *Eur. Phys. J.* **C 20** 479.
8. Muccifora V. for the HERMES Collaboration (2001) hep-ex/0106088.
9. HERMES Collaboration : Ackerstaff K. (1998) *Nucl. Instr. Meth.* **A 417** 230.
10. Guo X.and Wang X.N. (2000) *Phys. Rev. Lett.* **85** 3591.
11. Wang E.. and Wang X.N. (2002) *Nucl. Phys.* **A 702** 238.
12. Accardi A. and Pirner H.J. (2002) nucl-th/0205008.
13. Kopeliovich B. *et al.* (1995) hep-ph/9511214.
14. Akopov N.Z. *et al.* (2002) hep-ph/0205123 and references therein.
15. Cronin J.W. (1975) *Phys. Rev.* **D 11** 3105.
16. Kopeliovich B. *et al.* (2002) hep-ph/0201010.

HERMES RESULTS ON THE GENERALISED GDH INTEGRALS

N. AKOPOV
Yerevan Physics Institute, Alikhanian Brs. st. 2,
Yerevan, 375036, Armenia
E-mail: akopov@mail.desy.de
FOR THE HERMES COLLABORATION

Abstract.
The data collected by HERMES with a deuterium target are presented together with previous measurements on the proton. This provides an unprecedented and complete measurement of the generalised GDH integrals for the proton, deuteron and neutron for $1.2 < Q^2 < 12.0$ GeV^2 and for photon–nucleon invariant mass squared W^2 ranging over $1 < W^2 < 45$ GeV^2, thus covering both the nucleon-resonance and the deep inelastic scattering regions.

1. Introduction

The **Gerasimov-Drell-Hearn (GDH) sum rule** [1] relates the anomalous contribution κ in the nucleon magnetic moment to an energy weighted integral of the difference of the nucleon's polarised total photoabsorption cross sections:

$$\int_{\nu_0}^{\infty} [\sigma_{1/2}(\nu) - \sigma_{3/2}(\nu)] \frac{d\nu}{\nu} = -\frac{2\pi^2 \alpha}{M^2} \kappa^2, \tag{1}$$

where $\sigma_{1/2(3/2)}$ are the photoabsorption cross sections for total helicity equal to $1/2(3/2)$, ν is the photon energy, ν_0 is the pion photoproduction energy threshold, M is the nucleon mass. The GDH sum rule holds for any type of target, i.e. it is valid for protons, neutrons or nuclei. For the proton ($\kappa_p = +1.79$) the GDH sum rule prediction is -204μb. For the neutron ($\kappa_n = -1.91$), the prediction is -233μb. The sum rule arises from the combination of a general principles of causality, unitarity, crossing symmetry, and Lorentz and gauge invariance.

Many years the direct test was not performed, due to the need for circularly polarised beam and longitudinally polarised target and a wide range of photon ener-

E. Steffens and R. Shanidze (eds.), Spin Structure of the Nucleon, 121–131.

gies has to be covered. Only recently the first results on GDH sum rule checking from MAMI($0.2 \text{ GeV} \leq \nu \leq 0.8 \text{ GeV}$) has been published [2]. Using extrapolations into the unmeasured regions, Eq. (1) for the proton seems to be satisfied within the experimental uncertainties. Further real–photon experiments are underway at various laboratories to extend the energy range of the measurements [3, 4].

The GDH integral can be generalised to non–zero photon virtuality Q^2 in terms of the helicity–dependent virtual–photon absorption cross sections $\sigma^{\Leftarrow}$ and $\sigma^{\Rightarrow}$ [5, 6]:

$$I_{GDH}(Q^2) = \int_{\nu_0}^{\infty} [\sigma^{\Leftarrow}(\nu, Q^2) - \sigma^{\Rightarrow}(\nu, Q^2)] \frac{d\nu}{\nu}. \tag{2}$$

The cross section difference appearing in the integrand is given by

$$\Delta\sigma = \sigma^{\Leftarrow} - \sigma^{\Rightarrow} = \frac{8\pi^2\alpha}{M_t K} \tilde{A}_1 F_1. \tag{3}$$

In terms of photon–nucleon (nucleus) helicity states this relation is valid for any target; in case of the deuteron it comprises a mixture of vector and tensor states. Here $\tilde{A}_1$ is the photon–nucleon (nucleus) helicity asymmetry, F_1 the unpolarised nucleon (nucleus) structure function, M_t is the nucleus mass and K the virtual photon flux factor. The generalization of the GDH sum rule is not unique. Theoreticians tend to prefer a finite Q^2 generalization of Eq.(1) with theoretically well defined Q^2-evolution properties [7]. We prefer Eq.(2,3) because the asymmetries $A_1(x, Q^2)$ and $A_2(x, Q^2)$ are the quantity directly measurable in an experiment, Gilman flux factor $K = \nu\sqrt{(1+\gamma^2)}$ could be also replaced by Hand notation $K = \nu(1-x)$ [8, 9]

When considering a nucleon target (spin $\frac{1}{2}$, mass M) the photon helicity asymmetry $\tilde{A}_1$ is identical to the longitudinal virtual–photon asymmetry A_1 and the generalised GDH integral can be written in terms of the spin structure functions g_1 and g_2 as:

$$I_{GDH} = \frac{8\pi^2\alpha}{M} \int_0^{x_0} \frac{g_1(x, Q^2) - \gamma^2 g_2(x, Q^2)}{K} \frac{dx}{x}, \tag{4}$$

where g_1 and g_2 are the polarised structure functions of the nucleon, $\gamma^2 = Q^2/\nu^2$, $x = Q^2/2M\nu$, $x_0 = Q^2/2M\nu_0$.

Examining the generalised GDH integral provides a way to study the transition from polarised real–photon absorption ($Q^2 = 0$) on the nucleon to polarised deep inelastic lepton scattering (DIS). In other words, it constitutes an observable that allows the study of the transition from the non–perturbative regime at low Q^2 to the perturbative regime at high Q^2. Since the generalised GDH integral is calculated for inelastic reactions, elastic scattering is excluded from its calculation.

As has been pointed out in Ref. [10], the elastic contribution to the photon cross section becomes the dominant one below $Q^2 \simeq 0.5$ GeV^2 and has to be taken into account when comparing with twist expansions of the first moment of the spin structure function g_1. In the kinematic region considered in this paper, elastic contributions are expected to be small.

Due to the relatively large Q^2-values considered in this paper (where $\gamma^2 \simeq 0$), the Eq. (4) simplifies to

$$I_{GDH}(Q^2) = \frac{16\pi^2\alpha}{Q^2}\Gamma_1. \qquad (5)$$

The first moment of the the spin structure function g_1, $\Gamma_1 = \int_0^1 g_1(x)dx$, is predicted to have at large Q^2 only a logarithmic Q^2 dependence from QCD evolution. Since for the proton $\Gamma_1^p > 0$ for higher Q^2, I_{GDH}^p must change sign as Q^2 approaches zero in order to reach the negative value predicted by the GDH sum rule at the real–photon point. For the neutron Γ_1^n is negative for all measured Q^2.

The Q^2–dependence of the generalised GDH integral can be studied separately in the DIS region, characterised by large photon–nucleon invariant mass squared $W^2 = M^2 + 2M\nu - Q^2$, and in the nucleon–resonance region where W^2 amounts to only a few GeV^2. Several experiments measure the generalised GDH integral at low and intermediate Q^2, but cover kinematically only the low–W^2 region [11, 12, 13]. On the other hand, the high–W^2 contribution to the generalised GDH integral is found to be sizeable and essential to any estimate of the total integral [14, 15]. Preliminary data from real–photon experiments at higher energies support this statement [3]. The kinematics of the HERMES experiment allow the study of the Q^2–development of the generalized GDH integral simultaneously in both the nucleon-resonance and DIS region.

2. The Hermes Experiment

HERMES data on the deuteron target were taken in 1998 to 2000 with a 27.57 GeV beam of longitudinally polarised positrons incident on a longitudinally polarised atomic Deuterium gas target internal to the HERA storage ring at DESY. Data on the proton were taken in 1997 using a longitudinally polarised atomic Hydrogen target. The lepton beam polarisation was measured continuously using Compton backscattering of circularly polarised laser light [16, 17]. The average beam polarisation for the deuteron (proton) data set was 0.55 (0.55) with a fractional systematic uncertainty of 2.0% (3.4%).

The HERMES polarised gas target [18] consists of polarised atomic D (H) confined in a storage cell. The nuclear polarisation of atoms and the atomic fraction are continuously measured with a Breit–Rabi polarimeter [20] and a target gas analyser [21], respectively. The average value of the target polarisation for the deuteron (proton) data was 0.88 (0.86) with a fractional systematic uncertainty of 3.5 (4.4)%. The luminosity was monitored by detecting Bhabha events using calorimeter detectors close to the beam pipe [22]. The integrated luminosity per nucleon of the deuteron (proton) data set was 111 pb^{-1} (70 pb^{-1}).

Scattered positrons, as well as coincident hadrons, were detected by the HER-MES spectrometer [23]. Positrons were distinguished from hadrons with an average efficiency of 99% and a hadron contamination of less than 1% using the information from an electromagnetic calorimeter, a transition–radiation detector, a preshower scintillation counter and a Cherenkov counter. The threshold Cherenkov detector used in the proton measurement was replaced by a Ring–imaging Cherenkov detector [24] for the data taking on the deuteron. Only the information on the scattered positron was used in this analysis.

3. Data Analysis

The kinematic cuts for the scattered positrons in the analysis were identical for both targets. The full range in W^2 ($1.0 < W^2 < 45$ GeV^2) was separated into nucleon resonance region ($1.0 < W^2 < 4.2$ GeV^2) and DIS region ($4.2 < W^2 < 45.0$ GeV^2). The Q^2-range $1.2 < Q^2 < 12.0$ GeV^2 was divided into six bins; the same binning as in the proton case was chosen for the analysis of the deuteron data and for the subsequent determination of I_{GDH}^n. After applying data quality criteria, 0.55 (0.13) million events on the deuteron (proton) in the nucleon-resonance region and 8.3 (1.4) million events in the DIS region were selected. The detailed description of the analysis procedure and treatment of systematic uncertainties is given in [14, 15, 25].

For all detected positrons, the angular resolution was better than 0.6 mrad, the momentum resolution (aside from Bremsstrahlung tails) better than 1.6% and the Q^2–resolution better than 2.2%. The additional amount of material after the RICH installation led to a slightly worse W^2–resolution of $\delta W^2 \approx 1.0$ GeV^2 for the deuteron as compared to the proton measurement ($\delta W^2 \approx 0.82$ GeV^2). Although these W^2–resolutions do not allow distinguishing individual nucleon resonances, the integral measurement in the nucleon-resonance region is not degraded.

The generalised GDH integral Eq. (2) can be re–written for any target in terms of the photon–target helicity asymmetry $\tilde{A}_1$ and the unpolarised structure function F_1:

$$I_{GDH}(Q^2) = \frac{8\pi^2\alpha}{M_t} \int_0^{x_0} \frac{\tilde{A}_1(x, Q^2) F_1(x, Q^2)}{K} \frac{dx}{x}, \tag{6}$$

The longitudinal cross–section asymmetry $\tilde{A}_1$ for the absorption of virtual photons was calculated from the measured cross section asymmetry $A_{\parallel}$ as

$$\tilde{A}_1 = \frac{A_{\parallel}}{D} - \eta\tilde{A}_2. \tag{7}$$

where D and η are factors [25] that depend on kinematic variables, the D quantity depends also on $R = \sigma_L/\sigma_T$, the ratio of the absorption cross sections for longitu-

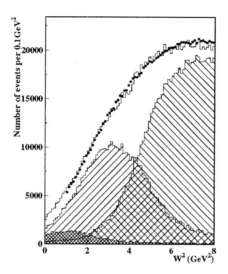

Figure 1.   Data – Monte Carlo comparison for the nucleon-resonance region as a function of W^2. The cross–hatched area represents the contribution from quasi–elastic scattering, while the lined areas show the contribution from the nucleon-resonance region (left) and from the DIS region (right). The solid line indicates the sum of all simulated events and compares favourably with the data points. The statistical uncertainties of the data are covered by the symbols.

dinal and transverse virtual photons. In the DIS region A_2 can be parametrised in a general form as $A_2 = cMx/\sqrt{Q^2}$, where c is a constant determined from a fit to the data given in Ref. [26, 27] as $c = 0.05\ (0.53)$ for the deuteron (proton). In the nucleon-resonance region no data are available for the deuteron and $A_2 = 0$ was chosen, while for the proton a constant value of $A_2 = 0.06 \pm 0.16$ was adopted as obtained from SLAC measurements at $Q^2 = 3\ \text{GeV}^2$ [26]. For spin $\frac{1}{2}$ targets the photon helicity asymmetry $\tilde{A}_1$ is identical to the longitudinal virtual photon asymmetry A_1 and $\tilde{A}_2$ is identical to A_2. The difference between these two asymmetries is relevant for the deuteron target only. Even here it is considered to be small in the kinematic region examined in this paper and hence will be neglected in the following. The extraction procedure for the cross–section asymmetry $A_\parallel$ is given in [15, 25].

Radiative effects for both targets were calculated using the codes described in Ref. [28]. They were found not to exceed 7% (4%) of the asymmetry A_1 for the deuteron (proton). On the integral level they do not exceed 2% and were included in the systematic uncertainty. The fraction of events smeared from the DIS to the nucleon-resonance region and vice versa is evaluated by a Monte Carlo simulation of both regions including radiative and detector effects. Smearing effects in the deep inelastic region have been evaluated for all targets following the procedures described in Ref. [15]. The events on the deuteron (proton) were simulated using the parametrisation of F_2 from Ref. [29] ([30]) for the DIS region, the elastic form factors from Ref. [31]([32]) and the parametrisation of F_2 in the nucleon-resonance region from Ref. [33] for both targets. Fig. 1 shows the distribution of experimental data as a function of W^2 in comparison with the simulated events on the deuteron. It is apparent that the shape of the simulated distribution agrees well with the data. Similar agreement has been found for the proton.

For the deuteron (proton) case, the relative contaminations from the elastic and deep inelastic region in the resonance region range from 15% (10%) to 3% (2%) and from 11% (7%) to 23% (16%) respectively, as Q^2 increases from 1.2 GeV^2 to 12.0 GeV^2. The fraction of events smeared from the resonance region to the deep inelastic region ranged from 2.9% (2.5%) to 0.5% (0.2%) respectively. Smearing from the elastic region to the DIS region can be neglected in the present experiment.

To evaluate the systematic uncertainty from smearing, two different assumptions on A_1 for the deuteron (proton) have been used: a polynomial representation $A_1 = -0.0307 + 0.92x - 0.28x^2$ (power law $A_1 = x^{0.727}$) that smoothly extends the DIS behaviour for the asymmetry into the resonance region [34]; and for both targets a step function ($A_1 = -0.5$ for $W^2 < 1.8$ GeV^2 and $A_1 = +1.0$ for 1.8 GeV$^2 < W^2 < 4.2$ GeV^2) that is suggested by the hypothesis of the possible dominance of the P_{33}–resonance at low W^2 and of the S_{11}–resonance at higher W^2 (see e.g. Ref. [35]). The combined systematic uncertainty from smearing and radiative effects does not exceed 14% (10%) for the deuteron (proton) data. In both cases, smearing gives by far the dominant contribution.

4. Results and Discussion

The GDH integrals for the deuteron and proton were evaluated following the procedure described in the previous section. The resonance region and the DIS region were treated separately. The large W^2–range covered by the HERMES experiment allows essentially the first experimental determination of the complete generalised GDH integral for the deuteron, proton and neutron. The GDH integral I_{GDH}^d for the deuteron was evaluated using Eq. (6) in the resonance region for $1.0 < W^2 < 4.2$ GeV^2 and in the DIS region for $4.2 < W^2 < 45.0$ GeV^2 . The unpolarised structure function $F_1 = F_2(1 + \gamma^2)/(2x(1 + R))$ was calculated from a modification of the parametrisation of F_2 given in Ref. [33] that accounts for nucleon resonance excitation assuming $R = \sigma_L/\sigma_T$ to be constant and equal to 0.18 in the whole W^2–range. The unpolarised structure function F_1^d for the deuteron in the DIS region was calculated following a parametrisation of F_2^d from Ref. [29]. In the same kinematic region R was chosen according to a fit in Ref. [36]. Note that due to cancellations between the R dependences of F_1 and D at low y the final result is affected by at most 2% by a particular choice of R. The W^2–dependence of the integrand F_1/K in the individual bins was fully accounted for in the inte-

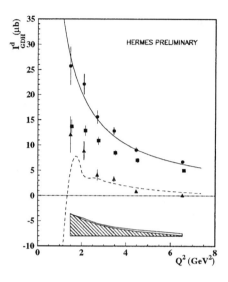

Figure 2. The generalised GDH integral I_{GDH}^d for the deuteron, shown as a function of Q^2 for the three different kinematic regions considered: resonance region (triangles), DIS region (squares), and full W^2–region including extrapolation to the unmeasured part (circles). The error bars show the statistical uncertainties. The upper curve is taken from Ref. [39], the lower curve represents a model for the resonance region from Ref. [38]. The systematic uncertainties of the full integral are given as a band, the hatched area inside represents the systematic uncertainty of the resonance region alone.

gration.

The extrapolation into the unmeasured region for $W^2 > 45$ GeV^2 was done using a multiple–Reggeon exchange parametrisation [37] for $\sigma_{\uparrow\downarrow} - \sigma_{\uparrow\uparrow}$ at high energy and ranged from -0.07 μb at $Q^2 = 1.5$ GeV^2 to 1.53 μb at $Q^2 = 6.5$ GeV^2. The corresponding contribution for the proton amounted to about 3.5 μb for all Q^2–bins.

The generalised GDH integrals for the deuteron data, calculated in the resonance region, in the DIS region and over the full W^2–range, are depicted in Fig. 2. The statistical and systematic uncertainties of the full I_{GDH} are clearly dominated by the uncertainties in the resonance region. They are particulary large due to the smallness of D and the large size of η accentuating the uncertainties in A_2, which amounts to 18%. The systematic uncertainty on A_2 in the DIS region does not contribute significantly. The systematic uncertainty for the extrapolation to the unmeasured region at high W^2 of 5% has been taken into account. Further sources of systematic uncertainties include the beam and target polarisations (5.5%), the spectrometer geometry (2.5%), the combined smearing and radiative effects (14%) and the knowledge of F_2 (5%). The total systematic uncertainty with respect to the total GDH integral values amounts to 18% at $Q^2 = 1.5$ GeV^2 to 7.5%

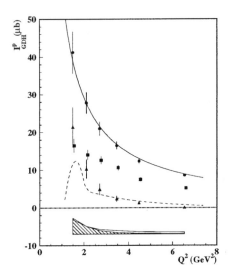

Figure 3. The generalised GDH integral I_{GDH}^p for the proton using the same notation as in Fig. 2. The upper curve is taken from Ref. [39], the lower curve follows the model of Ref. [38]. The data were published earlier in Ref. [15]. Note that the band representing the systematic uncertainties is given for the convention followed in this paper.

at $Q^2 = 6.5$ GeV^2. For the systematic errors of the resonance and DIS regions independent sources of systematic uncertainty were added in quadrature, while the systematic uncertainty stemming from smearing effects and the knowledge of F_2 were added linearly.

Following the formalism described in Ref. [40], the GDH integral I_{GDH}^n for the neutron was calculated from the GDH results on the deuteron I_{GDH}^d, as obtained in this analysis, and those on the proton I_{GDH}^p as presented in Ref. [15]:

$$I_{GDH}^n = \frac{I_{GDH}^d}{1 - 1.5\omega_d} - I_{GDH}^p. \tag{8}$$

Here $\omega_d = 0.058 \pm 0.010$ [41] is the probability of the deuteron to be in a D–state. It has been shown in Ref. [40] that although the uncertainties in the structure functions being integrated may be large, the resulting uncertainty for the integral does not exceed 3%.

The final results for the proton and for the neutron, as presented in Fig. 3 and Fig. 4, show that the contribution of the resonance region to the full generalised GDH integral is small for $Q^2 > 3$ GeV^2 and the contribution from the DIS region remains sizeable down to the lowest measured Q^2 values.

All data for the full generalised GDH integrals on the deuteron, proton and neutron are shown together in Fig. 5. They compare favourably to a prediction [39] based on the Q^2–evolution of the two polarised structure functions g_1 and g_2 without consideration of any explicit nucleon–resonance contribution. In the

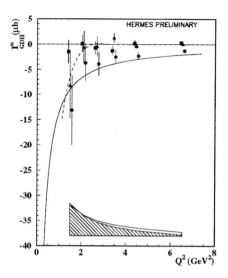

Figure 4. The generalised GDH integral I^n_{GDH} for the neutron obtained from the deuteron and proton using the same notation as in Fig. 2 for the symbols and theoretical curves.

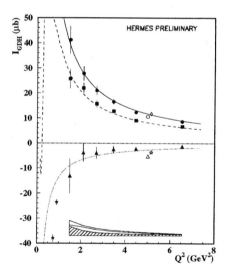

Figure 5. The Q^2-dependence of the generalised GDH integrals, calculated over the full W^2-region, for the deuteron (squares), proton (circles) and neutron (triangles). The latter was obtained from the deuteron and proton data. The curves shown are the predictions for the different targets according to Ref. [39]. The error bars represent the statistical uncertainties. The bands represent the systematic uncertainties (open: neutron, lined: deuteron, cross–hatched: proton). The open symbols represent the measurements from Ref. [26](left) and Ref. [27](right) on proton and neutron, respectively. The stars represent the two highest Q^2 bins of the neutron measurement from Ref. [13]

neutron case, the poorer knowledge of input data for this model leads to a larger uncertainty and a worse description of the data compared to the proton case. No sign change, as it would be required for the generalised GDH integral on the proton and the deuteron to meet the GDH sum rule prediction at $Q^2 = 0$, is observed in the measured range. Preliminary data from Ref. [11] and a recent theoretical evaluation indicate that this sign change happens at a value of Q^2 lower than the range considered in this analysis [7].

5. Conclusion

- The generalised Gerasimov-Drell-Hearn integral for the deuteron is experimentally measured for the first time in both the resonance and DIS regions for the Q^2 range from 1.2 to 12.0 GeV^2.
- Combining the deuteron data set with the already published data for the proton the GDH integral for the neutron was calculated in the same kinematic regions.
- Complete measurements of the GDH integrals for the proton, deuteron and neutron are available. In all three cases above $Q^2=3$ GeV^2 the DIS contribution to the GDH integral is dominant.
- At large Q^2 the measured values agree with data of the first moment of the spin structure functions g_1(p,n). Also the proton-neutron difference is in agreement with the BSR prediction within the total errors.
- The Q^2-dependence of the total GDH integral over the whole measured Q^2 range shows no significant modification to $1/Q^2$ behaviour due to resonances or higher-twist effects.

References

1. S.B. Gerasimov, Sov. J. Nucl. Phys. **2** (1966) 430; S.D. Drell and A.C. Hearn, Phys. Rev. Lett. **16** (1966) 908.
2. J.Ahrens et al., Phys. Rev. Lett. 87(2001), 022003
3. GDH collaboration, G. Zeitler, πN–Newsletter **16** (2002) 311.
4. V. Ghazikhanian et al., SLAC–Proposal E–159 (2000); D.Sober et al., CEBAF PR–91–15 (1991).
5. R. Pantförder, PhD Thesis, Universität Bonn (1998), BONN-IR-98-06, hep-ph/9805434 and reference therein.
6. D. Drechsel, S.S. Kamalov and L. Tiator, Phys. Rev. **D 63** (2001), 114010.
7. X. Ji, C.W. Kao and J. Osborne, Phys.Rev. **D 61** (2000) 074003 ; X. Ji and J. Osborne, J. Phys. **G 27** (2001) 127.
8. F.J. Gilman, Phys. Rev. **167** (1968) 1365.
9. L. N. Hand, Phys. Rev. **129** (1963) 1834.
10. X. Ji and W. Melnitchouk, Phys. Rev. **D 56** (1997) R1.
11. V. Burkert et al., CEBAF PR-91-23, 1991; S. Kuhn et al., CEBAF PR-93-09, 1993; J.P. Chen et al., TJANF PR-97-110, 1997.
12. CLAS Collaboration, R. Fatemi et al., Published in "Osaka 2000, Spin physics", 402.
13. M. Amarian et al., arXiv:nucl-ex/0205020
14. HERMES Collaboration, K. Ackerstaff et al., Phys. Lett. **B 444** (1998) 531.

15. HERMES Collaboration, A. Airapetian *et al.*, Phys. Lett. **B 494** (2000) 1.
16. D.P. Barber *et al.*, Phys. Lett. **B 343** (1997) 436.
17. M. Beckmann *et al.*, Nucl. Instr. Meth. **A 479** (2002) 334.
18. J. Stewart, Proc. of the Workshop on Polarised gas targets and polarised beams, edited by R.J. Holt and M.A. Miller, Urbana-Champaign, USA, AIP Conf. Proc. **421** (1997) 69.
19. F. Stock *et al.*, Nucl. Instr. and Meth. **A 343** (1994) 334.
20. C. Baumgarten *et al.*, Nucl. Instr. and Meth.**A 482** (2002) 606.
21. M. C. Simani *et al.*, submitted to Nucl. Instr. and Meth. **A**.
22. Th. Benisch *et al.*, Nucl. Instr. and Meth. **A 471** (2001) 314.
23. HERMES Collaboration, K. Ackerstaff *et al.*, Nucl. Instr. and Meth. **A 417** (1998) 230.
24. N. Akopov et al., Nucl. Instr. and Meth. **A 479** (2002) 511.
25. HERMES Collaboration, A. Airapetian *et al.*, Phys. Lett. **B 442** (1998) 484.
26. E–143 Collaboration, K. Abe *et al.*, Phys. Rev. **D 58** (1998) 112003.
27. E–155 Collaboration, P. L. Anthony *et al.*, Phys. Lett. **B 493** (2000) 19; E–155 Collaboration, P. L. Anthony *et al.*, Phys. Lett. **B 463** (1999) 339.
28. I.V. Akushevich and N.M. Shumeiko, J. Phys. **G 20** (1994) 513; I. Akushevich *et al.*, Comput. Phys. Commun. **104** (1997) 201.
29. NMC Collaboration, M. Arneodo *et al.*, Nucl. Phys. **B 371** (1992) 3.
30. NMC Collaboration, M. Arneodo *et al.*, Phys. Lett. **B 36 4** (1995) 107.
31. S. Stein *et al.*, Phys. Rev. **D 12** (1975) 1884 and references therein.
32. S. I. Bilen'kaya *et al.*, Zh. Eksp. Teor. Fiz. Pis'ma **19** (1974) 613.
33. A. Bodek, Phys. Rev. **D 8** (1973) 2331.
34. A. P. Nagaitsev *et al.*, JINR Rapid Communications, July 1995, N3(71)-95,59.
35. J. Edelmann, G. Piller, N. Kaiser and W. Weise, Nucl. Phys. **A 665** (2000) 125.
36. L.W. Whitlow *et al.*, Phys. Lett. **B 250** (1990) 193.
37. N. Bianchi and E. Thomas, Phys. Lett. **B 450** (1999) 439.
38. I. G. Aznauryan, Phys. of At. Nucl. **58** (1995) 1014 and private communication.
39. J. Soffer and O.V. Teryaev, Phys. Rev. **D 51** (1995) 25; J. Soffer and O.V. Teryaev, Phys. Rev. Lett. **70** (1993) 3373.
40. C. Ciofi degli Atti *et al.*, Phys. Lett. **B 376** (1996) 309.
41. M. Lacombe *et al.*, Phys. Lett. **B 101**(1981) 139.

SINGLE SPIN ASYMMETRY IN HEAVY QUARK PHOTOPRODUCTION (AND DECAY) AS A TEST OF PQCD

N.YA.IVANOV
Yerevan Physics Institute
Alikhanian Br.2, 375036 Yerevan, Armenia

1. Introduction

In the framework of perturbative QCD, the basic spin-averaged characteristics of heavy flavor hadro-, photo- and electroproduction are known exactly up to the next-to-leading order (NLO). During the last ten years, these NLO results have been widely used for a phenomenological description of available data (for a review see [1]). At the same time, the key question still remains open: How to test the applicability of QCD at fixed order to the heavy quark production? The problem is twofold. On the one hand, the NLO corrections are large; they increase the leading order (LO) predictions for both charm and bottom production cross sections approximately by a factor of two. For this reason, one could expect that higher-order corrections, as well as nonperturbative contributions, can be essential in these processes, especially for the c-quark case. On the other hand, it is very difficult to compare directly, without additional assumptions, pQCD predictions for spin-averaged cross sections with experimental data because of a high sensivity of the theoretical calculations to standard uncertainties in the input QCD parameters: m_Q, the factorization and renormalization scales, μ_F and μ_R, Λ_{QCD} and the parton distribution functions [2, 3].

In recent years, the role of higher-order corrections has been extensively investigated in the framework of the soft gluon resummation formalism [4, 5, 6]. Unfortunately, formally resummed cross sections are ill-defined due to the Landau pole contribution, and numerical predictions for the heavy quark production cross sections can depend significantly on the choice of resummation prescription [7].

For this reason, it is of special interest to study those observables which are well defined in pQCD. An nontrivial example of such an observable is proposed in [8, 9], where the charm and bottom production by linearly

E. Steffens and R. Shanidze (eds.), Spin Structure of the Nucleon, 133–140.

polarized photons,

$$\gamma^{\uparrow} + N \to Q + X[\overline{Q}], \tag{1}$$

was considered[1]. In particular, the single spin asymmetry (SSA) parameter, $A_Q(p_{QT})$, which measures the parallel-perpendicular asymmetry in the quark azimuthal distribution,

$$\frac{d^2\sigma_Q}{dp_{QT}d\varphi_Q}(p_{QT}, \varphi_Q) = \frac{1}{2\pi} \frac{d\sigma_Q^{\mathrm{unp}}}{dp_{QT}}(p_{QT}) [1 + A_Q(p_{QT})\mathcal{P}_\gamma \cos 2\varphi_Q], \tag{2}$$

has been calculated. In (2) $\frac{d\sigma_Q^{\mathrm{unp}}}{dp_{QT}}$ is the unpolarized cross section, $\mathcal{P}_\gamma$ is the degree of linear polarization of the incident photon beam and φ_Q is the angle between the beam polarization direction and the observed quark transverse momentum, p_{QT}. The following remarkable properties of the SSA, $A_Q(p_{QT})$, have been observed [8]:

- The azimuthal asymmetry (2) is of leading twist; in a wide kinematical region, it is predicted to be about 0.2 for both charm and bottom quark production.
- At energies sufficiently above the production threshold, the LO predictions for $A_Q(p_{QT})$ are insensitive (to within few percent) to uncertainties in the QCD input parameters.
- Nonperturbative corrections to the b-quark azimuthal asymmetry are negligible. Because of the smallness of the c-quark mass, the analogous corrections to $A_c(p_{QT})$ are larger; they are of the order of 20%.

In Ref. [9], radiative corrections to the φ-dependent cross section (2) have been investigated in the soft-gluon approximation. Calculations [9] indicate a high perturbative stability of the pQCD predictions for A_Q. In particular,

- At the next-to-leading logarithmic (NLL) level, the NLO and NNLO predictions for A_Q affect the LO results by less than 1% and 2%, respectively.
- Computations of the higher order contributions (up to the 6th order in α_s) to the NLL accuracy lead only to a few percent corrections to the Born result for A_Q. This implies that large soft-gluon contributions to the spin-dependent and unpolarized cross sections cancel each other with a good accuracy.

So, contrary to the the production cross sections, the single spin asymmetry in heavy flavor photoproduction is an observable quantitatively well

---

[1]The well known examples are the shapes of differential cross sections of heavy flavor production which are sufficiently stable under radiative corrections.

defined in pQCD: it is stable, both parametricaly and perturbatively, and insensitive to nonperturbative corrections. Measurements of this asymmetry would provide an ideal test of the conventional parton model based on pQCD.

Concerning the experimental aspects, the azimuthal asymmetry in charm photoproduction can be measured at SLAC where a coherent bremsstrahlung beam of linearly polarized photons with energies up to 40 GeV will be available soon [10]. In the planned E160 and E161 experiments, the charm production will be investigated using the inclusive spectra of the decay lepton:

$$\gamma^\uparrow + N \to c + X[\bar{c}] \to \mu^+ + X. \tag{3}$$

In this paper, we analyze a possibility to measure the SSA in heavy quark photoproduction using the decay lepton spectra. We calculate the SSA in the decay lepton azimuthal distribution:

$$\frac{d^2\sigma_\ell}{dp_{\ell T}d\varphi_\ell}(p_{\ell T}, \varphi_\ell) = \frac{1}{2\pi}\frac{d\sigma_\ell^{unp}}{dp_{\ell T}}(p_{\ell T})\left[1 + A_\ell(p_{\ell T})\mathcal{P}_\gamma \cos 2\varphi_\ell\right], \tag{4}$$

where φ_ℓ is the angle between the photon polarization direction and the decay lepton transverse momentum, $p_{\ell T}$. Our main results can be formulated as follows [11]:

- The SSA transferred from the decaying c-quark to the decay muon is large in the SLAC kinematics; the ratio $A_\ell(p_T)/A_c(p_T)$ is about 90% for $p_T > 1$ GeV.
- pQCD predictions for $A_\ell(p_{\ell T})$ are also stable, both perturbatively and parametricaly.
- Nonperturbative corrections to $A_\ell(p_{\ell T})$ due to the gluon transverse motion in the target and the c-quark fragmentation are small; they are about 10% for $p_{\ell T} > 1$ GeV.
- The SSA (4) depends weekly on theoretical uncertainties in the charm semileptonic decays[2], $c \to \ell^+ \nu_\ell X_q$ $(q = d, s)$. In particular,
  - Contrary to the the production cross sections, the asymmetry $A_\ell(p_{\ell T})$ is practically insensitive to the unobserved stange quark mass, m_s, for $p_{\ell T} > 1$ GeV.
  - The bound state effects due to the Fermi motion of the c-quark inside the D-meson have only a small impact on $A_\ell(p_{\ell T})$, in practically the whole region of $p_{\ell T}$.

So, we conclude that the SSA in the decay lepton azimuthal distribution (4) is also well defined in the framework of perturbation theory and can be used as a good test of pQCD applicability to heavy flavor production.

---

[2]For a review see Ref. [12].

2. pQCD Predictions for SSA

2.1. LO RESULTS

At the Born level, the only partonic subprocess which is responsible for the reaction (3) is the heavy quark production in the photon-gluon fusion,

$$\gamma^{\uparrow}(k_\gamma) + g(k_g) \to Q(p_Q) + \overline{Q}(p_{\overline{Q}}) \to \ell(p_\ell) + \nu_\ell + q + \overline{Q}, \qquad (5)$$

with subsequent decay $c \to \ell^+\nu_\ell q$ $(q = d, s)$ in the charm case and $b \to \ell^-\overline{\nu}_\ell q$ $(q = u, c)$ in the bottom one. To calculate distributions of final particles appearing in a process of production and subsequent decay, it is useful to adopt the narrow-width approximation,

$$\frac{1}{\left(p_Q^2 - m_Q^2\right)^2 + \Gamma_Q^2 m_Q^2} \to \frac{\pi}{\Gamma_Q m_Q} \delta\left(p_Q^2 - m_Q^2\right), \qquad (6)$$

with Γ_Q the total width of the heavy quark. Corrections to this approximation are negligibly small in both charm and bottom cases since they have a relative size $\mathcal{O}(\Gamma_Q/m_Q)$.

In the case of the linearly polarized photon, the heavy quark produced in the reaction (5) is unpolarized. For this reason, the single-inclusive cross section for the decay lepton production in (5) is a simple convolution:

$$E_\ell \frac{d^3\hat{\sigma}_\ell}{d^3p_\ell}(\vec{p}_\ell) = \frac{1}{\Gamma_Q} \int \frac{d^3p_Q}{E_Q} \frac{E_Q d^3\hat{\sigma}_Q}{d^3p_Q}(\vec{p}_Q) \frac{E_\ell d^3\Gamma_{sl}}{d^3p_\ell}(p_\ell \cdot p_Q). \qquad (7)$$

The leading order predictions for the φ_Q-dependent cross section of heavy flavor production,

$$\begin{aligned}
\frac{E_Q d^3\hat{\sigma}_Q}{d^3p_Q}(\vec{p}_Q) &\equiv \frac{2s d^3\hat{\sigma}_Q}{du_1 dt_1 d\varphi_Q}(s, t_1, u_1, \varphi_Q) \\
&= \frac{1}{\pi s} \left[B_Q(s, t_1, u_1) + \Delta B_Q(s, t_1, u_1) \mathcal{P}_\gamma \cos 2\varphi_Q \right], \qquad (8)
\end{aligned}$$

are given in [8]. Radiative corrections to the cross section (8) was investigated in the soft gluon approximation in Ref. [9].

At the tree level, the invariant width of the semileptonic decay $c \to \ell^+\nu_\ell q$ $(q = d, s)$ can be written as

$$\frac{E_\ell d^3\Gamma_{sl}}{d^3p_\ell}(x) \equiv I_{sl}(x) = \frac{G_F^2 m_Q^3}{(2\pi)^4} |V_{CKM}|^2 \frac{x(1 - x - \delta^2)^2}{1 - x} \qquad (9)$$

Here V_{CKM} denotes the corresponding element of the Cabbibo-Kobayashi–Maskawa matrix, G_F is the Fermi constant, $x = 2(p_\ell \cdot p_Q)/m_Q^2$ and $\delta = m_q/m_Q$.

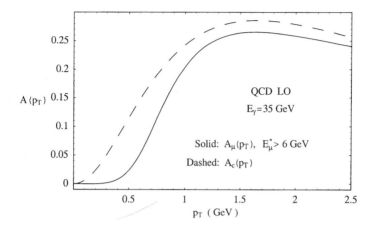

Figure 1. Comparison of the QCD LO pedictions for $A_\mu(p_T)$ and $A_c(p_T)$.

Let us discuss the hadron level pQCD predictions for the asymmetry in azimuthal distribution of the decay lepton. In this paper, we will consider only the charm photoproduction at the SLAC energy $E_\gamma \approx 35$ GeV, $E_\gamma = (S - m_N^2)/2m_N$. Unless otherwise stated, the CTEQ5M [13] parametrization of the gluon distribution function is used. The default value of the charm quark mass is $m_c = 1.5$ GeV.

Our calculations of the quantities $A_\mu(p_T)$ and $A_c(p_T)$ are given in Fig.1 by solid and dashed lines, respectively. One can see that the asymmetry transferred from the decaying c-quark to the decay muon is large in the SLAC kinematics; the ratio $A_\mu(p_T)/A_c(p_T)$ is about 90% for $p_T > 1$ GeV. Note that $p_T \equiv p_{QT}$ when we consider the heavy quark production and $p_T \equiv p_{\ell T}$ when the quantity $A_\mu(p_{\ell T})$ is discussed.

We have analyzed also the dependence of the SSA in the lepton distribution on the unobserved strange quark mass, m_s. Our analysis shows that the LO predictions for $A_\mu(p_T)$ are practically independent of $\delta = m_s/m_c$ at sufficiently large $p_T > 1$ GeV. For more details see Ref. [11].

2.2. RADIATIVE CORRECTIONS

We have computed both spin-dependent and unpolarized differential distributions (8) of the heavy-quark photoproduction at NLO to the next-to-leading logarithmic accuracy. The NLO corrections to the width of the heavy quark semileptonic decays are known exactly [14, 15]. We have found that radiative corrections to the leptonic SSA, $A_\mu(p_T)$, in the reaction (3) are of the order of (1-2)% in the SLAC kinematics.

Two main reasons are responsible for perturbative stability of the quantity $A_\mu(p_T)$. First, radiative corrections to the SSA in heavy quark production are small [9]. Second, the ratio $I_{\rm sl}^{\rm NLO}(x)/I_{\rm sl}^{\rm Born}(x)$ is a constant practically at all x, except for a narrow endpoint region $x \approx 1$ [15]. (Note that $I_{\rm sl}^{\rm Born}(x)$ is the LO invariant width of the semileptonic decay $c \to \ell^+ \nu_\ell X_q$ given by (9) while $I_{\rm sl}^{\rm NLO}(x)$ is the corresponding NLO one.)

3. Nonperturbative Contributions

Let us discuss how the pQCD predictions for single spin asymmetry are affected by nonperturbative contributions due to the intrinsic transverse motion of the gluon and the hadronization of the produced heavy quark. Because of the low c-quark mass, these contributions are especially important in the description of the cross section for charmed particle production [1]. At the same time, our analysis shows that nonperturbative corrections to the single spin asymmetry are not large.

Hadronization effects in heavy flavor production are usually modeled with the help of the Peterson fragmentation function [16],

$$D(y) = \frac{a_\varepsilon}{y\left[1 - 1/y - \varepsilon/(1-y)\right]^2},\tag{10}$$

where a_ε is a normalization factor and $\varepsilon_D = 0.06$ in the case of a D-meson production.

Our calculations of the asymmetry in a D-meson production at LO with and without the Peterson fragmentation effect are presented in Fig.2 by dotted and solid curves, respectively. It is seen that at $p_{DT} \geq 1$ GeV the fragmentation corrections to $A_c(p_T)$ are less than 10%.

Analogous corrections to the asymmetry in the decay lepton azimuthal distribution, $A_\mu(p_T)$, are given in Fig.3. One can see that the effect of the fragmentation function (10) is practically negligible in the whole region of $p_{\ell T}$.

To introduce k_T degrees of freedom, $\vec{k}_g \simeq z\vec{k}_N + \vec{k}_T$, one extends the integral over the parton distribution function to the k_T-space,

$$dz g(z, \mu_F) \to dz d^2 k_T f\left(\vec{k}_T\right) g(z, \mu_F).\tag{11}$$

The transverse momentum distribution, $f\left(\vec{k}_T\right)$, is usually taken to be a Gaussian:

$$f\left(\vec{k}_T\right) = \frac{e^{-k_T^2/\langle k_T^2\rangle}}{\pi\langle k_T^2\rangle}.\tag{12}$$

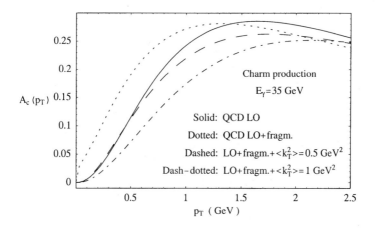

Figure 2.   SSA in a D-meson production; the QCD LO predictions with and without the inclusion of the k_T smearing and Peterson fragmentation effects.

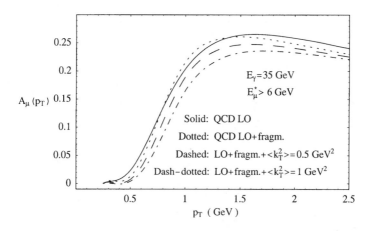

Figure 3.   SSA, $A_\mu(p_T)$, in the decay lepton distribution; the QCD LO predictions with and without the inclusion of the k_T smearing and Peterson fragmentation effects.

Values of the k_T-kick corrections to the asymmetry in the charm production, $A_c(p_T)$, are shown in Fig.2 by dashed ($\langle k_T^2 \rangle = 0.5$ GeV^2) and
dash-dotted ($\langle k_T^2 \rangle = 1$ GeV^2) curves. One can see that k_T-smearing is important only in the region of relatively low $p_{QT} \le m_c$. Note also that the
fragmentation and k_T-kick effects practically cancel each other in the case

of $\langle k_T^2 \rangle = 0.5$ GeV^2.

Corresponding calculations for the case of the lepton asymmetry are presented in Fig.3. It is seen that $A_\mu(p_T)$ is affected by k_T-corrections systematically smaller than $A_c(p_T)$.

4. Conclusion

In this paper we analyze the possibility to measure the SSA in open charm photoproduction in the E160/E161 experiments at SLAC where a coherent bremsstrahlung beam of linearly polarized photons with energies up to 40 GeV will be available soon. In these experiments, the charm production will be investigated with the help of inclusive spectra of the secondary leptons. The SSA transferred from the decaying c-quark to the decay muon is predicted to be large for SLAC kinematics; the ratio $A_\ell(p_T)/A_c(p_T)$ is about 90% at $p_T > 1$ GeV. Our calculations show that the SSA in decay lepton distribution preserves all remarkable properties of the SSA in heavy flavor production: it is stable, both perturbatively and parametricaly, and practically insensitive to nonperturbative contributions due to the gluon transverse motion in the target and the heavy quark fragmentation. We have also found that the QCD predictions for $A_\ell(p_T)$ depend weekly on theoretical uncertainties in the charm semileptonic decays. We conclude that measurements of $A_\ell(p_T)$ in the E160/E161 experiments would provide a good test of pQCD applicability to open charm production.

Acknowledgements. We would like to thank I.I. Balitsky, S.J. Brodsky, L. Dixon, A. Kotzinian, A.E. Kuraev, A.G. Oganesian, M.E. Peskin, A.V. Radyushkin and J. Tjon for useful discussions.

References

1. S.Frixione, M.L.Mangano, P.Nason and G.Ridolfi, hep-ph/9702287, published in "Heavy Flavours II", eds. A.J. Buras and M. Lindner, Advanced Series on Directions in High Energy Physics (World Scientific Publishing Co., Singapore, 1998).
2. M.L.Mangano, P.Nason and G.Ridolfi, Nucl. Phys. **B373** (1992), 295.
3. S.Frixione, M.L.Mangano, P.Nason and G.Ridolfi, Nucl. Phys. **B412** (1994), 225.
4. E.Laenen, J.Smith and W.L.van Neerven, Nucl. Phys. **B369** (1992), 543.
5. E.L.Berger and H.Contopanagos, Phys. Rev. **D 54** (1996), 3085.
6. S.Catani, M.L.Mangano, P.Nason and L. Trentadue, Nucl. Phys. **B478** (1996), 273.
7. N.Kidonakis, Phys. Rev. **D64** (2001), 014009.
8. N.Ya.Ivanov, A.Capella and A.B.Kaidalov, Nucl. Phys. **B586** (2000), 382.
9. N.Ya.Ivanov, Nucl. Phys. **B615** (2001), 266.
10. SLAC E161, (2000), http://www.slac.stanford.edu/exp/e160.
11. N.Ya.Ivanov et al., in preparation.
12. B.Blok, R.Dikeman and M.Shifman, Phys. Rev. **D51** (1995), 6167.
13. H.L.Lai et al., Eur. Phys. J. **C12** (2000), 375.
14. A.Ali and E.Pietarien, Nucl. Phys. **B154** (1979), 519.
15. M.Jeżabek and J.H.Kühn, Nucl. Phys. **B320** (1989), 20.
16. C.Peterson, D.Schlatter, I.Schmitt and P.Zerwas, Phys. Rev. **D 27**(1983), 105.

TESLA POLARIMETERS

V.GHARIBYAN, N. MEYNERS, K. P. SCHÜLER

DESY, Deutsches Elektronen Synchrotron, Hamburg, Germany

Abstract. We describe a study of high-energy Compton beam polarimeters for the future e^+e^- linear collider machine TESLA. A segment of the beam delivery system has been identified, which is aligned with the e^+e^- collision axis and which has a suitable configuration for high-quality beam polarization measurements. The laser envisaged for the polarimeter is similar to an existing facility at DESY. It delivers very short pulses in the 10 ps, $10 - 100$ μJ regime and operates with a pattern that matches the pulse and bunch structure of TESLA. This will permit very fast and accurate measurements and an expeditious tune-up of the spin manipulators at the low-energy end of the linac. Electron detection in the multi-event regime will be the principle operating mode of the polarimeter. Other possible operating modes include photon detection and single-event detection for calibration purposes. We expect an overall precision of $\Delta P/P \sim 0.5\%$ for the measurement of the beam polarization.

1. Introduction and Overview

A full exploitation of the physics potential of TESLA must aim to employ polarized electron and positron beams with a high degree of longitudinal polarization at full intensity. The technology of polarized electron sources of the strained GaAs type is well established [1, 2] and TESLA is therefore likely to deliver a state of the art polarized electron beam with about 80% polarization from the very beginning. The development of suitable beam sources of polarized positrons, based on the undulator method [3, 4] is still in its infancy but may soon be started with real beam tests at SLAC [5]. Equally important to the generation of high beam polarization will be its precise measurement and control over the full range of planned beam energies (45.6, 250, 400 GeV). The quantity of basic interest is the longitudinal

141

spin polarization of the two beams at the interaction point. Since a precise polarization measurement at the detector IP itself is difficult, the point of measurement should be chosen such that beam transport and beam-beam interaction effects are either negligible or small and well quantified. Other important factors relate to the level of radiation backgrounds and to the technical infrastructure and accessibility of a chosen site. The concept

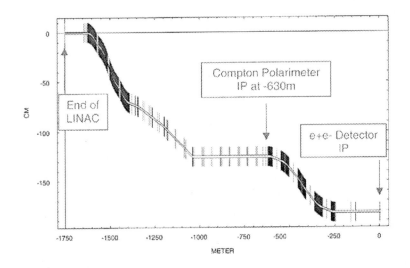

Figure 1. TESLA Beam Delivery System

of the polarimeter that we propose for TESLA is based on the well established laser backscattering method, as it was already envisaged in the TESLA CDR [2]. The proposed location of the Compton IP, where the laser beam crosses the electron or positron beam, is 630 meters upstream of the center of the e^+e^- detector, near the end of a long straight section of the beam delivery system (BDS), see Fig. 1. This part of the beamline is foreseen for general beam diagnosis and is also well suited for high quality beam polarization measurements.

Although the polarization vector experiences large rotations (due to the g-2 effect) as the beam traverses the bends of the BDS, the beam and spin directions at the chosen polarimeter site are precisely aligned, except for a parallel offset, with those at the e^+e^- interaction point. A polarization measurement at the proposed upstream location will therefore provide a genuine determination of the quantity of interest, as long as beam-beam effects are negligible or correctable. This is indeed the case. We estimate the beam-beam induced depolarization at TESLA to be 0.5%.

Fig. 2 shows a layout of the Compton Polarimeter. The laser beam crosses the electron or positron beam with a small crossing angle of 10 mrad at z = -630 m, just upstream of a train of ten C-type dipole magnets (BFCH0) which bend the beam horizontally by 0.77 mrad. The Compton scattered electrons are momentum analyzed in the field of the dipoles and detected with a segmented 14 channel gas Cerenkov counter. An optional calorimetric photon detector can also be employed further downstream.

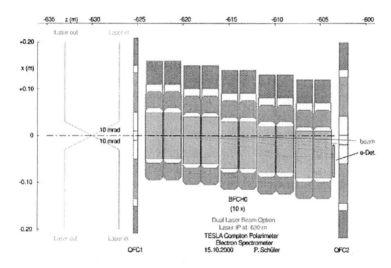

Figure 2. Layout of the Compton Polarimeter

The laser system that we envisage for the polarimeter should be similar to the laser configuration that has been developed by Max Born Institute for the Tesla Test Facility (TTF) photo injector gun at DESY [6]. This laser can be pulsed with a pattern that matches the peculiar pulse and bunch structure of TESLA. In this way it is possible to achieve very high luminosity, typically six orders of magnitude higher than with continuous lasers of comparable average power.

The statistics of Compton produced events is very high to the point where statistical errors will not matter in comparison with systematic errors. We expect a performance similar to the SLD Compton polarimeter at SLAC [7], with an overall precision of $\Delta P/P \sim 0.5\%$ for the measurement of the beam polarization.

2. General Considerations

The spin motion of a deflected electron or positron beam in a transverse magnetic field follows from the familiar Thomas-Larmor expression

$$\theta^{spin} = \gamma \frac{g-2}{2} \theta^{orbit} = \frac{E_0}{0.44065\ GeV} \theta^{orbit} \qquad (1)$$

where θ^{orbit} and θ^{spin} are the orbit and spin deflection angles, E_0 is the beam energy, $\gamma = E_0/m$, and $(g-2)/2$ is the famous g-factor anomaly of the magnetic moment of the electron.

In order to guarantee that the polarization measurement $\Delta P/P$ at the chosen polarimeter site does not suffer from systematic misalignments of the beam direction, we will postulate the following alignment tolerances

$$\Delta P/P \le 0.1\% \ \longrightarrow\ \Delta\theta^{spin} \le 45\,mrad \ \longrightarrow\ \Delta\theta^{orbit} \le 50\,\mu rad$$

where θ denotes the polar angle. The beam direction at the polarimeter site should therefore be aligned with the collision axis at the e^+e^- interaction point to within $50\mu rad$.

The strong beam-beam interaction at the collider IP will diffuse the angular spread of the beam. In Table 1 we have listed the rms values of the orbital angular spread of the disrupted beams at TESLA as obtained by O. Napoly. From the orbital rms values we have determined the associated rms spin distribution angles which are listed in Table 1. Based on these numbers, we estimate the overall depolarization of the spent beam to be $\Delta P/P \simeq 1 - cos(139mrad) = 1\%$, independent of beam energy. Assuming that the beam-beam interaction proceeds in a symmetric fashion upstream and downstream from the IP, we estimate the effective depolarization of the beam before the IP to be half of the overall effect, i.e. 0.5%.

	$\Delta\theta_x^{orbit}(rms)$ (μrad)	$\Delta\theta_y^{orbit}(rms)$ (μrad)	$\Delta\theta_x^{spin}(rms)$ $(mrad)$	$\Delta\theta_y^{spin}(rms)$ $(mrad)$
250 GeV	245	27	139	15
400 GeV	153	17	139	15

TABLE 1. Disrupted beam rms angular spreads of orbit and spin angles.

3. Compton Polarimeter

The Compton kinematics are characterized by the dimensionless variable

$$x = \frac{4E_0\omega_0}{m^2} \cos^2(\theta_0/2) \qquad (2)$$

where E_0 is the initial electron energy, ω_0 is the initial photon energy, θ_0

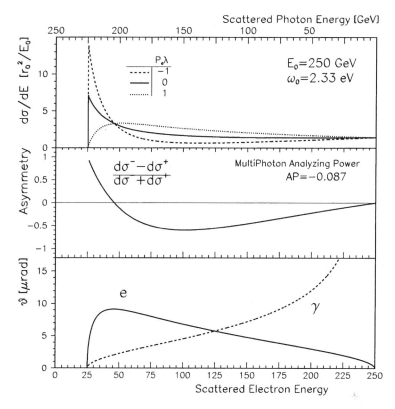

Figure 3. Energy spectra (top), spin asymmetry (middle) and scattering angles (bottom) of Compton scattered electrons and photons, for a beam energy of 250 GeV and a green laser.

is the crossing angle between the electron beam and the laser, m is the electron mass.

The energy spectra, the associated spin asymmetry and the scattering angles of the Compton scattered electrons and photons are shown in Fig. 3 for a beam energy of 250 GeV and a green laser (2.33 eV). For much higher or lower beam energies, it will be advantageous to change the wavelength of the laser. The multi-photon analyzing power A_p is also indicated in this figure.

The longitudinal polarization of the electron beam is determined from the asymmetry of two measurements of Compton scattering with parallel and antiparallel spin configurations of the interacting electron and laser beams.

For the TESLA Compton polarimeter, we plan to employ electron detection in the multi-event regime as the principle detection method. We will, however, reserve the multi-photon detection method as an option, es-

configuration	E_0 (GeV)	ω_0 (eV)	$<P_L>$ (W)	J (μJ)	$\mathcal{L}$ ($10^{32} cm^{-2} s^{-1}$)
TESLA-500	250	2.33	0.5	35	1.5
TESLA-800	400	1.165	1.0	71	6.0
Giga-Z	45.6	4.66	0.2	14	0.2

TABLE 2. Reference parameters for statistical tables.

bin #	min x_d (mm)	max x_d (mm)	E/E_0 low	E/E_0 high	Analyzing Power	Stat. Weight	$<d\sigma/dE>dE$ (mbarn)	Rate (MHz)
3	-75	-80	0.100	0.107	0.927	0.355	3.35	0.503
4	-70	-75	0.107	0.115	0.812	0.297	3.79	0.568
5	-65	-70	0.115	0.123	0.687	0.232	3.92	0.588
6	-60	-65	0.123	0.134	0.554	0.165	4.14	0.621
7	-55	-60	0.134	0.146	0.415	0.099	4.37	0.655
8	-50	-55	0.146	0.161	0.268	0.044	4.70	0.705
9	-45	-50	0.161	0.178	0.114	0.008	5.10	0.765
10	-40	-45	0.178	0.201	-0.044	0.001	5.57	0.835
11	-35	-40	0.201	0.229	-0.203	0.026	6.28	0.943
12	-30	-35	0.229	0.268	-0.355	0.075	7.25	1.087
13	-25	-30	0.268	0.321	-0.489	0.133	8.74	1.311
14	-20	-25	0.321	0.401	-0.577	0.176	11.28	1.692
all	-20	-90	0.089	0.401			68.49	10.273
Statistical Error for $\Delta t = 1$ second:				$\Delta P/P = 0.89 \cdot 10^{-3}$				

TABLE 3. Event rates and statistical error for TESLA-500.

pecially for TESLA operation at the Z-pole. Furthermore, we would like to point out that it can be very useful for calibration purposes to operate occasionally in the single-event regime, either with reduced pulse power of the laser or even with cw lasers. For the determination of event rates and statistical errors, we will use the reference parameters listed in Table 2, where P_L (J) is the laser average(pulse) intensity and $\mathcal{L}$ is luminosity. In order to be consistent with the cross sections in Fig. 3, we list the wavelengths for a Nd:YAG laser. The wavelengths of the Nd:YLF laser are only slightly different. Not explicitly listed are the crossing angle $\theta_0 = 10\ mrad$ and the size of the laser focus $\sigma_{x\gamma} = \sigma_{y\gamma} = 50\ \mu m$, which are assumed to be common for all configurations.

    The Table 3 lists the binned cross sections and event rates in the electron detector (E and x_d are the scattered electron energy and position) for the

	e^+/e^- beam	laser beam
energy	250 GeV	2.3 eV
charge or energy/bunch	$2 \cdot 10^{10}$	35 μJ
bunches/sec	14100	14100
bunch length σ_t	1.3 ps	10 ps
average current(power)	45 μA	0.5 W
$\sigma_x \cdot \sigma_y$ (μm)	$10 \cdot 1$	$50 \cdot 50$
beam crossing angle	10 mrad	
luminosity	$1.5 \cdot 10^{32} cm^{-2} s^{-1}$	
cross section	$0.136 \cdot 10^{-24} cm^2$	
detected events/sec	$1.0 \cdot 10^7$	
detected events/bunch	$0.7 \cdot 10^3$	
$\Delta P/P$ stat. error/sec	negligible	
$\Delta P/P$ syst. error	$\sim 0.5\%$	

TABLE 4. Compton Polarimeter Parameters at 250 GeV

first reference configuration of Table 2. As these events are bunched and recorded as analog signals at the bunch crossing frequency, there is no problem with apparently high rates, as we do not actually count individual events.

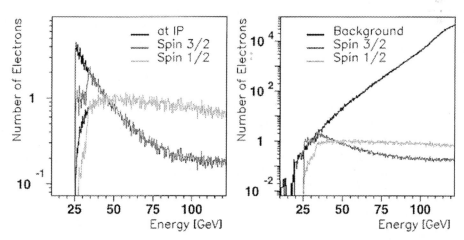

Figure 4. Simulated per bunch spectra of scattered electrons for two laser helicities tracked down to $z = 65$ m. Superimposed: spectra at $z = 0.8$ m (left) and background at $z = 65$ m (right).

The analyzing power A_i for each bin and the associated statistical weights w_i for a beam polarization P $= 0.80$ are also given in Table 3.

Furthermore, we list the statistical errors $\Delta P/P$ of the beam polarization for a measurement duration Δt of 1 second. We conclude from these numbers that genuine statistical errors originating from the Compton event statistics will be exceedingly small and likely negligible in comparison with systematic effects. We expect an overall precision of $\Delta P/P \sim 0.5\%$ for the measurement of the beam polarization. As an example, Table 4 gives typical polarimeter parameters for TESLA-500. The performance is similar for other energy regimes of TESLA. More details are given in [12].

We have also considered the possibility of downstream polarimeter locations, which would in principle permit to investigate beam-beam effects experimentally. However, disrupted beam polarimetry appears very difficult because of severe background from energy-degraded beam electrons as shown in Figure 3 for 250 GeV electrons, green laser and nominal beam extraction parameters with reduced 10 mm collimation at $z = 18$ m.

References

1. R. Alley, H. Aoyagi, J. Clendenin, J. Frisch, C. Garden, E. Hoyt, R. Kirby, L. Klaisner, A. Kulikov, R. Miller, G. Mulhollan, C. Prescott, P. Saez, D. Schultz, H. Tang, J. Turner, *The Stanford Linear Accelerator Polarized Electron Source*, Nucl. Instr. and Meth. A365 (1995) 1.
2. R. Brinkmann et al., *Conceptual Design Report of a 500 GeV e^+e^- Linear Collider with integrated X-ray Laser Facility*, DESY-1997-048, 1997.
3. V.E. Balakin and A.A. Mikhaillichenko, *The Conversion System for Obtaining High Polarized Electrons and Positrons*, Preprint INP 79-85, 1979.
4. K. Flöttmann, *Investigation Toward the Development of Polarized and Unpolarized High Intensity Positron Sources for Linear Colliders*, DESY-93-161, 1993.
5. SLAC-Proposal-E166, October 2002.
6. I. Will, P. Nickles and W. Sandner, *A Laser System for the TESLA Photo-Injector*, internal design study, Max-Born-Institut, Berlin, 1994.
7. M. Woods, *The Scanning Compton Polarimeter for the SLD Experiment*, SLAC-PUB-7319, 1996; 12th Int. Symposium on High-Energy Spin Physics (Spin96), NIKHEF, Amsterdam, Proceeding eds. C.W. de Jager et al., World Scientific, 1997, p. 843.
8. I.F. Ginzburg, G.L. Kotkin, V.G. Serbo, and V.I. Telnov, Nucl. Instr. and Meth. 205 (1983) 47.
9. W.Y. Tsai, L.L. DeRaad, and K.A. Milton, Phys. Rev. D6 (1972) 1428; K.A. Milton, W.Y. Tsai, and L.L. DeRaad, Phys. Rev. D6 (1972) 1411.
10. H. Veltman, *Radiative corrections to polarized Compton scattering*, Phys. Rev. D40 (1989) 2810; Erratum Phys. Rev. D42 (1990) 1856.
11. M. L. Swartz, *A Complete Order-α^3 Calculation of the Cross Section for Polarized Compton Scattering*, Phys. Rev. D58:014010 (1998); hep-ph/9711447; SLAC-PUB-7701, 1997.
12. V.Gharibyan, N.Meyners, P.Schuler, *The TESLA Compton Polarimeter*, DESY LC-DET-2001-047, 2001; http://www.desy.de/~lcnotes/.

HERA TRANSVERSE POLARIMETER ABSOLUTE SCALE AND ERROR BY RISE-TIME CALIBRATION

V.GHARIBYAN

DESY, Deutsches Elektronen Synchrotron, Hamburg, Germany

Yerevan Physics Institute, Yerevan, Armenia

AND

K. P. SCHÜLER

DESY, Deutsches Elektronen Synchrotron, Hamburg, Germany

Abstract.

We give the results of an analysis of some 18 rise-time calibrations which are based on data collected in 1996/97. Such measurements are used to determine the absolute polarization scale of the transverse electron beam polarimeter (TPOL) at HERA. The results of the 1996/97 calibrations are found to be in good agreement with earlier calibrations of the TPOL performed in 1994 with errors of 1.2% and 1.1%. Based on these calibrations and a comparison with measurements from the longitudinal polarimeter (LPOL) at HERA carried out over a two-months period in 2000, we obtain a mean LPOL/TPOL ratio of 1.018. Both polarimeters are found to agree with each other within their overall errors of about 2% each.

1. Introduction

Two polarimeters are employed at the HERA ep storage ring to measure the polarization of its 27.5 GeV electron or positron beam. Both instruments are laser backscattering Compton devices. The TPOL polarimeter measures the transverse beam polarization by detecting the associated angular anisotropy of the backscattered Compton photons. The original configuration of this instrument has been covered in considerable detail [1, 2, 3, 4, 5] and recent upgrades are described in [6]. The LPOL polarimeter measures the longitudinal beam polarization between the spin rotators at the HER-

E. Steffens and R. Shanidze (eds.), Spin Structure of the Nucleon, 149–156.

MES experiment by detecting an asymmetry in the energy spectra of the Compton photons [7].

In this paper, we will present the results of an analysis of rise-time calibration data which were collected in 1996/97. Such measurements are used to determine the absolute polarization scale of the TPOL polarimeter. These results will be compared with earlier TPOL calibrations obtained in 1994 and with recent cross calibrations with the LPOL polarimeter.

2. The rise-time calibration method

Electrons or positrons are injected unpolarized at 12 GeV into the HERA storage ring and are subsequently ramped to the nominal beam energy of 27.5 GeV. Transverse polarization evolves then naturally through the spin flip driven by synchrotron radiation (the Sokolov-Ternov effect [8]) with an exponential time dependence

$$P(t) \; = \; P^{\infty} \left(1 - \exp\left(-t/\tau\right)\right) \tag{1}$$

For a circular machine with a perfectly flat orbit the spin vector of the positrons (electrons) will be exactly parallel (antiparallel) to the direction of the guide field and the theoretical maximum of the polarization has been calculated to be $P^{\infty}_{ST} = 8/(5\sqrt{3}) = 92.4\%$, with an associated rise-time constant

$$\tau_{ST} \; = \; P^{\infty}_{ST} \, \frac{m_e |\rho|^3}{r_e \hbar \gamma^5} \tag{2}$$

where γ is the Lorentz factor, ρ is the radius of curvature of the orbit and the other symbols have the usual meaning.

For rings such as HERA with the spin rotators needed to get longitudinal polarization at experiments and/or reversed horizontal bends, P^{∞}_{ST} can be reduced substantially below 92.4% and τ_{ST} can be modified too, see Table 1. Synchrotron radiation also causes depolarization which competes with the Sokolov-Ternov effect with the result that the equilibrium polarization is reduced even further. Moreover the depolarization is strongly enhanced by the presence of the small but non-vanishing misalignments of the magnetic elements and the resulting vertical orbit distortions which are typically 1 mm rms.

These effects are treated in the formalism of Derbenev and Kondratenko [9] which has been summarized in [10]. Then the equilibrium polarization and the time constant can be written as

$$P^{\infty} \; = \; -\frac{8}{5\sqrt{3}} \frac{\oint ds \, \langle \, |\rho(s)|^{-3} \, \hat{b} \cdot (\hat{n} - \partial\hat{n}/\partial\delta) \, \rangle_s}{\oint ds \, \langle \, |\rho(s)|^{-3} \, [1 - \frac{2}{9}(\hat{n} \cdot \hat{s})^2 + \frac{11}{18}(\partial\hat{n}/\partial\delta)^2] \, \rangle_s} \tag{3}$$

$$\tau = \frac{8}{5\sqrt{3}} \frac{m_e}{r_e \hbar \gamma^5} C \frac{1}{\oint ds \, \langle \, |\rho(s)|^{-3} \, [1 - \frac{2}{9}(\hat{n} \cdot \hat{s})^2 + \frac{11}{18}(\partial \hat{n}/\partial \delta)^2] \, \rangle_s} \qquad (4)$$

where the unit vector $\hat{n}$ describes the polarization direction which is a function of the machine azimuth s and the phase space coordinate $\vec{u} = (x, p_x, y, p_y, z, \delta)$, the unit vectors $\hat{b}$ and $\hat{s}$ describe the magnetic field orientation and the direction of motion, and C is the circumference of the machine. The angular brackets $\langle \, \rangle_s$ denote an average over phase space at azimuth s. The term with $(\partial \hat{n}/\partial \delta)^2$ accounts for the radiative depolarization due to photon-induced longitudinal recoils and the term with $\partial \hat{n}/\partial \delta$ in the numerator of 3 arises from the dependence of the radiation power on the spin orientation.

These expressions can be summarized in the scaling relation

$$\frac{P^\infty}{(P_{ST}^\infty + \Delta)} = \frac{\tau}{\tau_{ST}} \qquad (5)$$

between the actually observed parameters P^∞ and τ of equation 1 and the theoretical values P_{ST}^∞ and τ_{ST} which are obtained by ignoring terms with $\partial \hat{n}/\partial \delta$ in equations 3 and 4. For the HERA machine at 27.5 GeV they take the values in Table 1 [11]:

HERA status	year	P_{ST}^∞	τ_{ST} (min)
flat	1994	0.915	36.7
non-flat	1996/97	0.891	36.0

TABLE 1. Input parameters for rise-time calibrations

The correction Δ in equation 5 results from the term $\partial \hat{n}/\partial \delta$ in the numerator of equation 3. In the case of a flat ring, Δ is negligible compared to P_{ST}^∞. However, for a non-flat HERA, namely with spin rotators activated at HERMES, Δ/P_{ST}^∞ remains small but can still be significant. Since the magnitude of Δ depends on the distortions of the machine, it is very difficult to calculate reliably [11].

The scaling relationship of equation 5 can be exploited to predict the expected equilibrium polarization P^∞ when we know the remaining quantities P_{ST}^∞, τ_{ST}, τ and Δ. Through comparison of the actually measured value of P^∞ with the predicted value from equation 5 we can therefore calibrate the polarimeter.

This is the essence of the rise-time method. It requires knowledge of the theoretical maximum values for the machine in the absence of depolarizing spin diffusion effects, i.e. P_{ST}^{∞} and τ_{ST} from equations 3 and 4, and a measurement of the actual rise-time τ from a fit of data to the functional form of equation 1. For a non-flat ring we also need to know Δ. However, by comparison with rise-time calibrations obtained earlier with a flat machine, we will be able to examine this term experimentally.

3. Experimental Procedure

In order to take rise-time calibration data which are free from major systematic effects it is essential that the build-up of polarization proceeds under conditions of an extremely stable machine performance since even minor operator adjustments in the course of a measurement may change the parameters of the functional form that describes the time dependence. For this reason rise-time calibrations require dedicated HERA operation, where the machine is brought to a very stable condition and is then only monitored without further operator invention. The beam polarization is then destroyed by the resonant depolarization technique by applying an rf field to a weak kicker magnet [12]. When the baseline polarization P_0 has been established the depolarizing rf is turned off at time $t = 0$ and the subsequent exponential rise is measured under completely quiet machine conditions. After one or two hours, the polarization can be destroyed again for another rise-time measurement and so forth.

In order to retain only data of high quality, we applied the following selection criteria: (a) stable machine and polarimeter conditions; (b) depolarizing rf frequency shifts by no more than 100 Hz (corresponding to a change in beam energy of about 1 MeV); in order to apply this test the beam needs to be depolarized before and after each rise-time measurement; (c) the depolarizer should be activated for about 10 minutes prior to $t = 0$ to establish a reliable baseline P_0. Of the rise-time data collected in 1996/97, altogether 18 curves out of a total of 25 survived these cuts. For the older measurements taken in 1994, 8 curves out of 14 could be retained [13].

4. Results and Errors

Figure 1 shows two examples of rise-time data obtained in 1997. The curves are fits to the following functional form, with τ, P_0 and K as free fitting parameters

$$K \cdot \left[P(t) - P_0 \right] = \frac{P_{ST}^{\infty}}{\tau_{ST}} \tau \left[1 - \exp\left(-t/\tau\right) \right] \qquad (6)$$

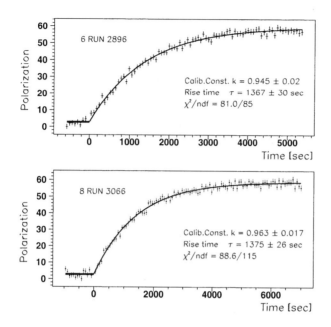

Figure 1.   Examples of rise-time calibration measurements

The appropriate input values for P_{ST}^{∞} and τ_{ST} were listed in table 1. The parameter K is the calibration factor for the polarization measurement of the polarimeter. The results of the 1996/97 rise-time fits are listed in table 2, together with the older calibration data from 1994 which had been reported earlier [13].

In table 3 we give the weighted mean value of K, $\bar{K} = \sum w_i K_i / \sum w_i$ with $w_i = (\Delta K_i)^{-2}$, the error of the mean value $\Delta \bar{K} = (\sum w_i)^{-1/2}$, and the associated $\chi^2/ndf = \sum w_i (K_i - \bar{K})^2/(N-1)$. Furthermore we give a rescaled error $\langle \Delta \bar{K} \rangle_{scaled} = (\chi^2/ndf)^{1/2} \cdot \Delta \bar{K}$ to account for the underestimation of the original errors indicated by $\chi^2/ndf > 1$. The scaling is equivalent to adding a common systematic error of size $(\Delta \bar{K})_{syst} = \Delta \bar{K} \cdot (\chi^2/ndf - 1)^{1/2}$ in quadrature, see [14] for an explanation of these procedures.

For a proper interpretation of the calibration results shown in table 3, it is important to understand that all TPOL polarization values have been scaled by a factor 0.946 since 1996 [13] in good agreement with the value of 0.951 given here. The subsequent 1996/97 relative re-calibration factor of 0.999 is therefore equivalent to an overall absolute factor of $0.999 \cdot 0.946 = 0.945$ in relation to the original polarization scale used prior to 1996.

The analysis of the 1994 rise-time data reported in [13] assigned an error of 0.032 to the determination of K. The error quoted is the rms σ of the nearly gaussian distribution of K-measurements. This rms σ value

Index No.	Year	Run No.	K	ΔK	τ (sec)	$\Delta\tau$ (sec)	χ^2/ndf
1	1994	2370	0.936	0.021			
2	1994	2442	0.976	0.042			
3	1994	2444	0.900	0.038			
7	1994	2482	0.960	0.030			
8	1994	2484	0.956	0.023			
9	1994	2486	0.955	0.028			
10	1994	2488	0.994	0.021			
11	1994	2492	0.902	0.026			
1	1996	5138	0.971	0.042	1255	55	0.644
2	1996	7702	1.089	0.028	1637	42	0.788
3	1996	8382	1.006	0.023	1328	31	0.765
4	1997	2778	0.996	0.042	1392	61	0.680
5	1997	2824	0.962	0.032	1342	47	0.828
6	1997	2896	0.945	0.020	1367	30	0.953
7	1997	3030	0.998	0.031	1396	49	0.763
8	1997	3066	0.963	0.017	1374	25	0.770
9	1997	3278	0.906	0.052	1136	65	0.701
10	1997	3316	0.999	0.043	1251	57	0.597
11	1997	3318	1.030	0.070	1254	90	0.516
14	1997	3669	0.933	0.039	873	37	0.821
15	1997	6654	1.038	0.047	1172	53	0.922
16	1997	6684	1.075	0.035	1362	49	0.784
17	1997	6686	1.062	0.038	1388	53	0.838
18	1997	6688	0.977	0.035	1265	49	0.702
24	1997	9610	1.056	0.026	1391	37	0.795
25	1997	9642	1.040	0.024	1463	36	0.507

TABLE 2. Calibration results from rise-time measurements

data set	$\bar{K}_{rel}$	$\bar{K}_{abs}$	$\Delta\bar{K}$	χ^2/ndf	$(\Delta\bar{K})_{syst}$	$(\Delta\bar{K})_{scaled}$
1994	0.951	0.951	0.009	1.509	0.006	0.011
1996/97	0.999	0.945	0.007	2.744	0.009	0.012

TABLE 3. Mean calibration factors and errors

describes the typical error of a single measurement of K. The error of the mean value of K is then $\sigma/\sqrt{N}$ and thus considerably smaller than 3%.

By comparing the calibrations of 1994 and 1996/97, we find excellent agreement within the given errors of 1.1 and 1.2%. Since the 96/97-calibrations were carried out with activated spin rotators at HERMES, in contrast to 1994 when the machine was flat, we can also set an experimental upper limit of about 1.5% on the correction Δ in equation 5 which accounts for the $\partial\hat{n}/\partial\delta$ term in the numerator of equation 3.

5. LPOL/TPOL comparison

As the magnitude of the polarization at any particular point in time is an invariant around the HERA ring, a comparison between the TPOL and LPOL polarimeters will provide a cross calibration of the instruments, as long as the spin points fully upright at the TPOL and longitudinal at the LPOL.

Although it is well known to the operators of the polarimeters and to the members of HERMES that the two instruments appeared to disagree occasionally with each other outside of their quoted errors, it is reassuring to demonstrate a large sample of measurements, covering two months of data taking in April/May of 2000, which exhibits a consistent and stable performance of both polarimeters. We have plotted the LPOL/TPOL ratio for this time period in figure 2 for six different integration times ranging from 1 to 25 minutes. We obtain a consistent mean ratio of 1.018 with an error of the mean $\sigma/\sqrt{N}$ of less than 0.001.

Since the statistical fluctuations are vanishing with higher averaging periods, one can estimate the systematic error of the TPOL measurement $(\Delta P/P)_{TPOL}$ from the observed rms $\sigma = 0.025$ of the 25 min distribution and the quoted LPOL systematic error $(\Delta P/P)_{LPOL} = 1.6\%$ [7]. Assuming uncorrelated errors for the two polarimeters, we obtain $(\Delta P/P)_{TPOL} = [\sigma^2 - (\Delta P/P)^2_{LPOL}]^{1/2} = 1.9\%$.

6. Acknowledgments

The rise-time calibration analysis presented in this article and the LPOL/TPOL comparison is based on data collected by the HERMES Polarimeter Group. We thank Desmond Barber for discussions on radiative polarization and depolarization mechanisms.

References

1. D.P. Barber et al., *The HERA polarimeter and the first observation of electron spin polarization at HERA*, Nucl. Instr. Meth. A329 (1993) 79.

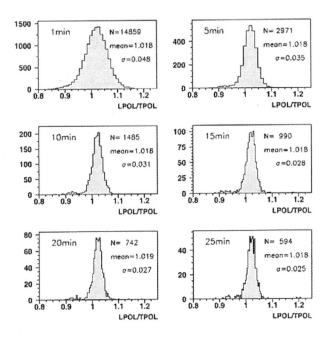

Figure 2.   LPOL/TPOL ratio measurements over a two-months period in 2000

2. D.P. Barber et al., *High Spin polarization at the HERA electron storage ring*, Nucl. Instr. Meth. A338 (1994) 166.

3. D.P. Barber et al., *The first achievement of longitudinal spin polarization in a high energy electron storage ring*, Phys. Lett. B434 (1995) 436.

4. D. Westphal et al., *Polarisationsmessungen am HERA Elektronenstrahl*, DESY F35D-93-04, diploma thesis, Univ. Hamburg, 1993.

5. K.P. Schüler, *Polarimetry at HERA*, Proc. 12th Int. Symp. on High Energy Spin Physics (SPIN 96) Amsterdam, eds. C.W. de Jaeger et al., World Scientific, 1997, p. 822.

6. K. Long, K.P. Schüler, *Polarisation measurements on $e^\pm$ beams*, NIM A, article in press.

7. M. Beckmann et al., *The Longitudinal Polarimeter at HERA*, Nucl. Instr. Meth. A479 (2002) 334.

8. A.A. Sokolov, I.M. Ternov, Sov. Phys. Doklady 8 (1964) 1203; I.M. Ternov, Yu.M. Loskutov, L.I. Korovina, Sov. Phys. JETP 14 (1962) 921.

9. Ya.S. Derbenev, A.M. Kondratenko, Sov. Phys. JETP 37 (1973) 968; S.R. Mane, Phys. Rev. A36 (1987) 105.

10. D.P. Barber, G. Ripken, *Radiative Polarization, Computer Algorithms and Spin Matching in Electron Storage Rings*, Handbook of Accelerator Physics and Engineering, eds. A.W. Chao and M. Tigner, 2nd edition, World Scientific, 2002.

11. D.P. Barber, private communications

12. F. Zetsche, *Use of Resonant Depolarization at the HERA Electron Ring*, Proc. 12th Int. Symp. on High Energy Spin Physics (SPIN 96) Amsterdam, eds. C.W. de Jaeger et al., World Scientific, 1997, p. 846.

13. S. Barrow et al., *Interim Report on the Measurement of the Positron Beam Polarization*, HERMES Polarimeter Note, Jan. 1996.

14. Particle Data Group, *Review of Particle Physics, Chapter 4.2 Averages and fits*, Eur. Phys. J. C3 (1998) p. 10.

THE TRANSVERSE POLARIZED TARGET OF THE HERMES EXPERIMENT

D. REGGIANI (FOR THE HERMES COLLABORATION)

Istituto Nazionale di Fisica Nucleare,
Laboratori Nazionali di Frascati, 00044 Frascati, Italy.

Dipartimento di Fisica, Università di Ferrara,
44100 Ferrara, Italy.

e-mail: reggiani@mail.desy.de

Abstract. The HERMES polarized hydrogen/deuterium internal target is operational since 1996 in the HERA lepton storage ring. In this contribution the target setup is described and an overview of the performance achieved during 2000 data taking is given. In 2001 the direction of the target holding field has been switched from longitudinal to transverse, with respect to the momentum of the lepton beam. The studies carried out in order to accomplish this result are presented. A critical aspect the new HERMES target is the possible nucleon depolarization caused by the interaction of the transient magnetic fields generated by the beam with the polarized nucleons of the target.

1. Introduction

HERMES is an inclusive and semi-inclusive DIS experiment which makes use of the longitudinally polarized e^+ or e^- beam provided by the HERA storage ring at DESY. HERMES is furnished with an internal polarized hydrogen or deuterium gaseous target combined with the storage cell technique [1]. The advantages of a gaseous target are the absence of dilution by spectator nuclei and the possibility of reversing the nuclear spin within a period of time of a few ms. Moreover, the use of the storage cell results in an increase of the target density by about two orders of magnitude with respect to a free jet target.

In its original implemetation, the HERMES target has been conceived in

E. Steffens and R. Shanidze (eds.), Spin Structure of the Nucleon, 157–166.

order to provide longitudinal nuclear polarization to the lepton beam momentum [2]. This setup has been kept from 1996 till the end of 2000. The growing interest for the measurement of the tranverse quark spin distribution has pushed the HERMES collaboration for the design of a tranversely polarized target. This led to the construction and installation of a new dipolar target magnet which provides a vertical holding field for the target spin. As a consequence of the interaction between the lepton beam and the tranverse magnetic field, technical issues as the bump correction and the emission of synchrotron radiation towards the HERMES spectrometer had to be taken into account. Further, the superimposition of the beam induced RF field and the target static tranverse field inside the storage cell could lead to some nuclear depolarizing resonances, which are much more difficult to avoid than in the longitudinal case.

A positive feasibility test performed in 1999 drove the collaboration to the decision of installing a tranversely polarized target, taking advantage of the one year HERA shutdownd between September 2000 and September 2001.

2. The HERMES target

The HERMES polarized target consists of an Atomic Beam Source (ABS) which produces a polarized jet of atomic hydrogen or deuterium and focuses it into a storage cell, a windowless elliptical tube through which the HERA electron or positron beam passes.

A sample of gas (ca. 5%) diffuses from the middle of the cell into a Breit Rabi Polarimeter (BRP), which measures the atomic polarization, and a Target Gas Analizer (TGA), which measures the atomic and the molecular content of the sample. A magnet sorrounding the storage cell provides an holding field defining the polarization axis and prevents spin relaxation via spin exchange or wall collisions by effectively decoupling the magnetic moments of electrons and nucleons.

The presently employed HERMES storage cell (fig. 1) consists of a 21 x 8.9 mm elliptical tube 400 mm long made from 75 micron thick alluminum. The polarized atomic hydrogen is injected through a 100 mm long 10 mm diameter tube and flows out of the cell through its open ends into the beam pipe and through a 100 mm long 5 mm diameter sample tube towards the diagnostic devices. Both injection and sample tubes are mounted at an angle of 30° with respect to the orizontal plane. A differential pumping system with a total pumping speed of 9000 l/s is used to evacuate the HERA lepton ring from the target gas. A bigger storage cell (29 x 9.8 mm cross section) has been used from 1996 till 1999. The current cell has been succesfully installed for the first time at the beginning of 2000 and has led to a raise of the target density by a factor 1.5.

Wall collisions can cause atomic depolarization and recombination [3, 4]. In order to inhibit as much as possible these effects, the HERMES cell is coated with Drifilm. During HERA beam operation, this coating can be partialy spoiled by synchrotron radiation. As a small amount of oxigen is coming from the dissociator, an additional water coating gradually covers the cell wall. This ice coverage has been found to compensate for the loss of the Drifilm surface.

A gaseous helium cooling system mounted on the cell support structure, in combination with a heater resistor placed inside the helium line, can lower the cell temperature down to about 35 K and keep it constant. This is important to increase the target thickness, which rises as $1/\sqrt{T}$, and, at the same time, to optimize the target performance by setting temperature values where the atoms are less sentitive to the depolarization and recombination mechanisms. Those values have been found to be 100 K for hydrogen and 60 K for deuterium.

The storage cell is sorrounded by an holding magnetic field provided by two different magnets: the longitudinal target applies a superconducting solenoid with a nominal axial field of 0.335 T and a uniformity of ±2%, while the transverse one employes a conventional dipole whose features are described in the last part of this proceeding. In the HERMES ABS (fig. 2) atomic hydrogen (or deuterium), produced by means of a radiofrequency dissociator, flows through an aluminum nozzle, a skimmer and a collimator, gets into a sextupolar magnetic system, which produces an electron polarized beam, and then into high frequency transition units, which transfer the

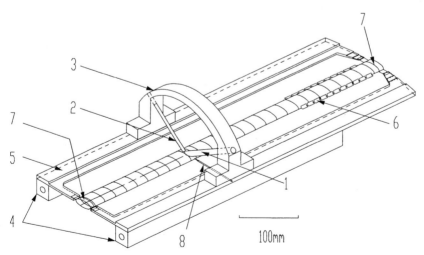

Figure 1.   The HERMES target cell. The different components are: 1) H/D injection tube; 2) Sample tube; 3) Support arch; 4) Cryogenic-cooling and support rails; 5) Support plates; 6) Cell extension with pumping apertures; 7) Cell end support collars; 8) Unpolarized gas feed capillary.

polarization to the nuclei. The polarized beam flows then into the storage cell through the injection tube.

The rf-dissociator (replaced only for the 2000 deuterium running period by a microwave) operates at a frequency of 13.56 MHz with a nominal power of 300 W. The input flux consists of 80 sccm of hydrogen together with a small amount of oxigen which improves the discharge properties. This leads to the formation of an ice coating on the nozzle surface, constanlty kept at 100 K during normal operation. In order to prevent blocking, the nozzle is periodically warmed up to room temperature with a tipical cycle between four and seven days for hydrogen. The sextuple magnets system is composed by radially segmented permanent magnets with pole tip fields up to 1.5 T. They focus atoms in the two (three) higher hyperfine levels and defocus the lower two (three) in the hydrogen (deuterium) case. Four adiabatic hyperfine transition units are used to interchange the atom state populations. Two units are placed in the so called ABS-appendix, just before the injection tube, while the other two are located between the sextuple subsystems. In the deuterim case all units are needed to supply nuclear positive and negative polarization, whereas for hydrogen only the appendix ones are requested for this pourpose. A powerful differential pumping system with a total nominal pumping speed of over 150000 l/s ensures a low attenuation of the beam caused by residual gas.

The HERMES ABS is capable of a throughput of $6.6 \cdot 10^{16}$ atoms/s for hydrogen injeting two hyperfine states and $5.0 \cdot 10^{16}$ atoms/s for deuterium with three states injected.

In the Breit-Rabi polarimeter [5] the sampled target gas encounters first two hyperfine transition units and then a sextuple magnets system. A flux composed by the two (three) higher energy hyperfine states of atomic hydrogen (deuterium) is focused into a quadrupole mass spectrometer (QMS) and detected by a Channeltron. A beam blocker placed inside the sextuple

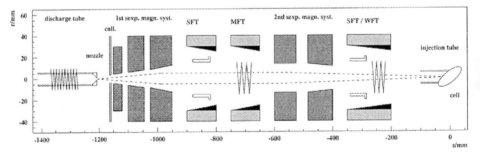

Figure 2. Schematic of the HERMES ABS. The dark grey elements represent the sextupolar magnets, while the light grey ones are the transition units. The strong field transitions (SFTs) have to be replaced, while the medium/weak field transitions (MFT/WFT) have to be retuned when swithing between hydrogen and deuterium.

system ensures that no atoms in the lower states can reach the QMS. The background measurement is carried out by using a chopper which periodically shuts the flux in front of the QMS. A differential pumping system keeps the pressure in the detector chamber at $1 \cdot 10^{-10}$ mbar.

During a given ABS injection status, the BRP transition units are operated in at least four (six) different modes in the hydrogen (deuterium) case. In this way, a number of signals equal to or bigger than the number of hyperfine levels can be collected. Knowing the efficiencies of the transitions units and the sextuple system relative transmissions for different hyperfine states, the four (six) hyperfine populations can be calculated. Finally, applying the knoledge of the target field intensity, the polarization of the sampled atomic beam is computed. During normal running conditions, the entire procedure lasts roughly 60 s for hydrogen and 90 s for deuterium.

The BRP calibration is carried out by operating the transition units in all possible ways. By doing that for different ABS injections modes, it is possible to collect a number of signals bigger than the number of unknown (efficiencies, transmission ratios and hyperfine populations) and therefore determine the transition units efficiencies and the sextuples transmission ratios.

The BRP sextuple system has been recently optimized by replacing its magnets. Due to this improvement, the current statistical uncertainty of a 60 s polarization measurement is less than 0.5 %. The systematic error is in the order of 1 %.

The BRP vacuum system is shared by the target gas analyzer, which measures the relative hydrogen (deuterium) flux of Mass 1 (2) and Mass 2 (4) exiting the target cell's sample tube. The TGA is mounted 7° off axis with respect to the BRP, in order not to interfere with the beam entering the polarimeter. The TGA detector follows the scheme of the BRP one: a chopper, a QMS and a Channeltron plus a pair of buffles for beam collimation. The pressure in the TGA QMS chamber is around $1 \cdot 10^{-9}$ mbar.

The TGA is calibrated by injecting a constant atomic flux into the target cell and varying the amount of recombination. This can happen either by changing the cell temperature or after a HERA beam loss near the HERMES experimental hall. The calibration constant can be determined by plotting the background corrected Mass 2 (4) versus Mass 1 (2) rates and fitting them with a straigh line. This value can be estimated with a 5 % accuracy.

The statistical error of a 60 s atomic fraction measurement is in the order of 2.5 %.

3. Overview on the 2000 target performance

During the period between January and September 2000 the HERMES target has continously run with deuterium. In the whole time range, the scattering chamber had no need to be opened and no bed events by the HERA positron beam occurred.

The average target polarization can be written as:

$$P^T = \alpha_0[\alpha_r + (1 - \alpha_r)\beta]P_a$$

where α_0 is the injected atomic fraction inside the cell, α_r represents the atomic fraction which survived recombination on the cell wall, P_a is the atomic polarization and β is the ratio between molecular and atomic polarization. As the sampled target gas is extracted from the center of the cell, none of those average quantities can be directly measured. The measurable ones are called:

$$\alpha_0^{TGA} = \frac{\phi_a + \phi_r}{\phi_a + \phi_r + \phi_{rg} + \phi_{bal}} \quad , \quad \alpha_r^{TGA} = \frac{\phi_a}{\phi_a + \phi_r} \quad , \quad P_a^{BRP}$$

where ϕ_a, ϕ_r, ϕ_{rg} and ϕ_{bal} are the normalized TGA hydrogen or deuterium fluxes belonging respectively to atoms, recombined moleculs (polarized), target chamber's residual hydrogen molecules (unpolarized) and ballistically injected molecules (unpolarized).

The cell averaged quantities α_r and P_a can be estimated by introducing the so called sampling corrections c_α and c_P [6]:

$$\alpha_r = c_\alpha \cdot \alpha_r^{TGA} \quad , \quad P_a = c_P \cdot P_a^{BRP}$$

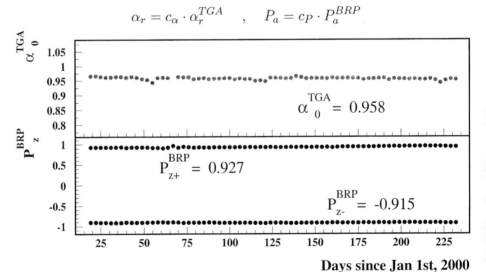

Figure 3.   The HERMES deuterium target performance in 2000. In the upper plot the initial atomic fraction α_0^{TGA} is shown, while in the lower one the positive and negative atomic nuclear polarizations are displayed.

Knowing α_r, also α_0 can be calculated, while β has not been directly measured yet.

In the particular case of no recombination and no depolarization inside the cell, $c_\alpha = c_P = 1$ and therefore:

$$\alpha_r = \alpha_r^{TGA} = 1 \quad , \quad P_a = P_a^{BRP}$$

This leads to a great semplification of the target polarization formula:

$$P^T = \alpha_0 \cdot P_a^{BRP}$$

The data analisys has shown that during the 2000 running period the HERMES polarized deuterium target behaved in an ideal way, in the sense that no evidence of atomic recombination or depolarization inside the storage cell has been found. Moreover, the ABS performance has been absolutely stable overall the data taking period (fig. 3). For these reasons average values could be used to describe the target nuclear positive and negative polarizations, obtaining the following results [7]:

$$P_{z+} = +0.851 \pm 0.031 \quad , \quad P_{z-} = -0.840 \pm 0.028$$

The estimated target thickness for two injected states is:

$$\rho\Delta z = 2.1 \cdot 10^{14} \; nucleons \cdot cm^{-2}$$

4. The transverse magnet design

The target magnet has two purposes: to determine the orientation of the target spin and to effectively decouple the magnetic moments of electrons and nucleons. This latter feature is of fundamental importance in order to achieve a high injected polarization and to preserve it from spin relaxation processes. Indeed, in the HERMES storage cell one can distinguish between different depolarization mechanisms: wall depolarization, taking place when an atom hits the cell wall, spin exchange, occuring when two atoms collide in the gas phase, and bunch field induced resonant depolarization, originating when the frequency of an rf-harmonic induced by the HERA e-beam matches the frequency difference between two different hyperfine states present in the cell. The first two processes can be prevented or at least limited by applying an adequately strong magnetic field, whereas a high field uniformity is required in order to advoid the third one. On the other hand, in the transverse target magnet case serius constraints on the maximum appliable field originate from the bending of the HERA electron beam passing through the vertical field and the consequent emission of synchrotron light towards the HERMES spectrometer.

The lower limit for the field intensity has been derived from the experience accumulated with the longitudinal target: in order to maintain the atomic polarization above 85 % and the total polarization relative uncertainty below 4.5 %, a field strength of around 300 mT is highly desirable (fig. 4).

The problem of the upper appliable field strength had to be studied carefully. An estimation has shown that the e-beam trajectory could be compensated up to 340 mT target field by using two correction magnets already existing upstream and downstream of the target, at a distance of z=-1270 mm and z=2189 mm respectively. At the nominal beam momentum of 27.5 GeV/c a beam displacement inside the storage cell $d=1.75$ mm has been estimated. The total bending angle $\Phi=2.2$ mrad would lead to a very narrow cone of emitted synchrotron radiation confined inside the electron beam pipe through the whole HERMES spectrometer length. With a beam current of 50 mA the total emitted power would be around 5 KW, a value absolutely compatible with the accelerator design.

The most critical requirement for the transverse field comes from the study of the possible beam induced depolarization. When the frequency difference between a pair of hyperfine states matches the frequency of one of the beam harmonics, resonace depolarization occurs with a probability proportional to the square of the beam current. In order to know for which values of the target field this mechanism can take place, one has to study the harmonic structure of the time dependent magnetic field induced by the 220 bunches of the electron beam. As the distance between two adjacent bunches is $\tau=96$ ns, the frequency spacing between two harmonics is given by $\nu=\frac{1}{\tau}=10.41$ MHz. Moreover, since the width of the gaussian shaped bunch is very narrow ($\sigma_t=37.7$ ps), a huge number of harmonics

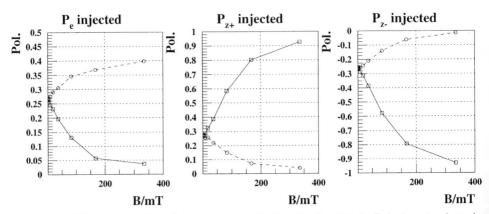

Figure 4.   The polarization of the atoms sampled by the Breit-Rabi Polarimeter plotted as function of the holding field in the hydrogen case. The open circles (squares) represent the electron (proton) polarization. The three graphs show different ABS injection modes: electron, nuclear positive and nuclear negative polarizations (from left to right).

with not negligible amplitude (more than 400 within a sigma of the Fourier spectrum) can contribute to induce the magnetic rf-field.

Depending on which pair of hyperfine states is involved, transitions are distinguished between π, occurring when the rf-field component is perpendicular to the static one, and σ, taking place when the two fields are parallel. Around the working point of the target magnet (300 to 340 mT), the π resonances are easily avoidable with a field uniformity of the percent level. This is unfortunately not the case for the σ, whose spacing is only 0.37 mT. All resonance conditions which can happen in the hydrogen case are shown in figure 5.

Beam induced depolarization in the HERMES target has already been observed in the longitudinal case [8] where, for geometrical reasons, only the π resonances can occur. On the countrary, no measurement of the σ resonances exists for the hydrogen case.

Taking into account the teoretical width of the σ resonances, the requirement for the target field uniformity has been fixed to the conservative value of $\Delta B = 0.14$ mT.

Due to geometrical constraints imposed by the HERMES setup, the design of a magnet which would meet all the listed requirements was not possible. After the construction, the field uniformity has been improved by shimming the magnet pole tips. At a field intensity of $B = 297$ mT, values of $\Delta B_z = 0.05$ mT, $\Delta B_y = 0.15$ mT, $\Delta B_x = 0.60$ mT have been acheaved, all measured overall the storage cell dimensions, where z is the logitudinal di-

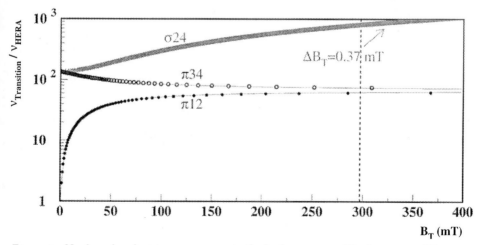

Figure 5. Nuclear depolarizing resonances in the hydrogen case. The frequency difference between pairs of hydrogen hyperfine states whose transitions would lead to a nuclear depolarization are plotted as function of the holding field. The frequency values are normalized for $\nu_{HERA} = 10.41$ MHz. The marks, representing the resonance condition, are clearly distinguishable for the π transitions, while they overlap with each other in the σ case.

rection, while y and x are the two transverse ones. Although not completely satisfactory, the magnet has been installed into the HERMES target in July 2001 and the whole experimental setup has been arranged in order to produce and measure polarized hydrogen. Due to the poor performance of the HERA machine after the startup in September 2001, a serius study of the effect of the field non-uniformity along x has not been possible up to now. A feasible solution making use of two correction coils embedded inside the storage cell support structure is currently under test.

5. Conclusion

The HERMES gaseous target has operated reliably providing high longitudinal polarization for hydrogen and deuterium from 1996 to 2000. The target performance was particularly remarkable during the 2000 deuterium runnig period, both for the considerable stability of the atomic beam source and for the absence of any depolarization or recombination process inside the storage cell.

Thanks to the design and construction of a new magnet, the HERMES target is now capable of providing transverse polarized hydrogen. Atomic nuclear polarization up to $P_a^{BRP}=85\%$ has been measured with a positron beam corrent up to $10\,\mathrm{mA}$. A higher beam current is needed in order to study the possible resonant induced depolarization effect and to decide whether a correction for the not perfect uniformity of the target magnet will be required.

References

1. C. Baumgarten et al., The Storage Cell of the Polarized H/D Internal Target of the HERMES Experiment at HERA, to appear in *Nucl. Instrum. and Meth. A*.
2. H. Kolster, Ph.D. Thesis, Ludwig-Maximilians Universitaet Muenchen, February 1998.
3. C. Baumgarten, Ph.D. Thesis, Ludwig-Maximilians Universitaet Muenchen, May 2000.
4. C. Baumgarten et al., Measurements of Atomic Recombination in the HERMES Polarized Hydrogen and Deuterium Storage Cell Target, to appear in *Nucl. Instrum. and Meth. A*.
5. C. Baumgarten et al., *Nucl. Instrum. and Meth. A*, **482 (2002)**, 606.
6. C. Baumagarten et al., *Eur. Phys. J. D* **18 (2002)**, 37.
7. The HERMES Target Group, The polarization of Hermes target and its error for the deuterium running in 2000, HERMES Internal, Note 2002.
8. K. Ackerstaff et al., *Phys. Rev. Lett.*, **82 (1999)** 1164-1168.

EXCLUSIVE MESON PRODUCTION AT HERMES

D.RYCKBOSCH, *ON BEHALF OF THE HERMES COLLABORATION*
University of Gent, Dept. of Subatomic Physics, Gent, Belgium

1. Introduction

The introduction of the framework of Generalized Parton Distributions over the past few years has revitalized the field of hard exclusive reactions. In these reactions, where a hard scale is introduced by the momentum transfer Q^2 $(= -(k - k')^2$ where $k(k')$ is the 4-momentum of the incoming (scattered) electron), the final state is particularly simple: basically it contains the nucleon in its ground state plus one or two emitted particles. The description of the reaction becomes very clear if the emitted particle is a real photon: Deeply Virtual Compton Scattering (DVCS). However, also the emission of single mesons is open to description in terms of GPDs.

In the present proceedings a short overview is given of recent results obtained by the HERMES experiment on the hard exclusive production of mesons.

2. The HERMES experiment

HERMES is a fixed target experiment[4] on the 27.6 GeV HERA-e ring. The electrons in the ring are transversally polarized by the Sokolov-Ternov effect, and longitudinal polarization is achieved by a spin rotator in front of the HERMES target. A second rotator behind the experiment restores the transverse polarization. Typical beam polarizations are between 50 and 60%. The beam traverses the HERMES target consisting of a 40 cm long windowless storage cell which confines the injected polarized target gas to the region around the beam. A small sample of the target gas is analyzed in a Breit-Rabi polarimeter. In 1996 and 1997 data were taken on a polarized proton target.

The scattered electrons, and the hadrons produced in the reaction, are detected in the HERMES spectrometer. This is a typical forward magnetic

E. Steffens and R. Shanidze (eds.), Spin Structure of the Nucleon, 167–172.

spectrometer with a dipole field strength of 1.3 T·m. A full description of
the spectrometer is given in ref.[4].

With the 27.6 GeV electron beam in HERA the kinematic range covered
by the HERMES experiment and relevant for the present analysis is $3.8 <
W < 6.5$ GeV, and $0.5 < Q^2 < 5$ GeV^2. This places HERMES in roughly
the same range of momentum transfer as the ZEUS and H1 experiments,
but at a much lower W.

3. Cross sections

The produced vector mesons are identified by determining the invariant
mass of the set of detected decay particles: $\rho^0 \to \pi^+\pi^-, \phi \to K^+K^-, \omega \to
\pi^+\pi^-\pi^0$. As an example the invariant mass spectrum for 3 pions detected
in HERMES is shown in fig.1.

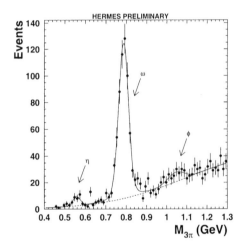

Figure 1.   The 3 pion invariant mass spectrum.

The exclusivity of the data sample is ensured by making a cut in the
excitation energy spectrum: $\Delta E = \frac{M_Y^2 - M^2}{2M} < 0.6$ GeV, where M_Y is the
invariant mass of the (undetected) final state. A typical example can be
found in ref.[5].

4. pQCD description

The exclusive production of mesons can be described in terms of the Generalized Parton Distributions that were introduced a few years ago. These
GPDs form a natural off-forward extension of the standard Parton Distribution Functions which are well determined from (semi-)inclusive DIS

reactions. They form a connection between the PDFs and the form-factors of hadrons. In fact, the framework of GPDs can in the case of vector meson production also be seen as an extension of the well known description of diffractive reactions at higher energies. In that case the reaction is assumed to proceed through the exchange of a colour-neutral pair of gluons (or the Pomeron) between the vector meson and the target nucleon. At lower energies the exchange is not of a pair of gluons but of quarks. This is then described in terms of GPDs.

The cleanest example of the appearance of GPDs in an exclusive reaction is that of DVCS[?]. In the case of meson production an extra unknown part of soft physics appears: the distribution amplitude of the meson. However, there are also advantages to the study of exclusive meson production. Firstly, they can act as helicity filters: production of vector mesons involves only the GPDs H and E, while scalar meson production is sensitive only to $\tilde{H}$ and $\tilde{E}$. In contrast to this all four GPDs appear in the description of DVCS. Secondly, the meson may obviously also act as a flavour filter.

5. Results and discussion

In figure 2 the total cross section for ϕ-production is shown as a function of W for different values of Q^2. Also shown in this plot are the cross sections determined by other experiments at higher and lower W. It is obvious that the same behaviour is seen over the entire energy range. The reason for this is that the quark exchange mechanism is suppressed in ϕ-production where it would involve s-quarks. Thus the same Pomeron exchange mechanism that prevails at higher energies is still dominant at the HERMES energy.

Using the W-dependence as found in fig.2 one can scale all cross sections to a common W value (of 75 GeV) and then study the Q^2 dependence. This is shown in fig.3. In this figure the cross section for vector meson production from the HERA collider experiments H1[2] and ZEUS[1] are compared to the HERMES data for ϕ- and ρ^0-production. It is clear that all cross sections seem to follow a kind of universal curve as function of Q^2, at least at the high W-values of the HERA collider experiments. The same behaviour is found at HERMES for the ϕ meson, but the ρ^0 cross section (and the ω cross section, not shown) is about a factor of 2 higher. This points to the importance of a reaction mechanism other than the 2-gluon exchange.

Recent calculations using Generalized Parton Distributions give a very reasonable description of the HERMES data. This is shown in fig.4 where the cross section for production of ϕ and ω vector mesons are compared to theoretical calculations[3].

A similar good agreement between theory and experiment is found in

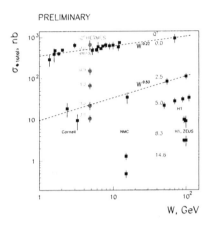

Figure 2. ϕ-production cross section as function of W for different values of Q^2.

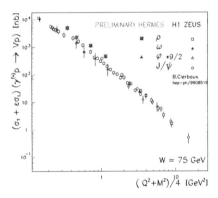

Figure 3. Vector meson production cross section as determined at H1 and ZEUS and HERMES experiments. All cross sections are scaled to a common W-value of 75 GeV.

the case of ρ^0 production[5]. For both the ω and ρ^0 the calculation involving quark exchange gives a good description of the data, confirming the importance of this reaction mechanism.

A further confirmation of this can be found in the ratio of cross sections. Figure 5 displays the ratio of ω to ρ^0 cross section as a function of the photon virtuality. It can be seen that this ratio exhibits very little dependence on Q^2. For the H1 and ZEUS results a ratio of about 1/9 is found, in accordance

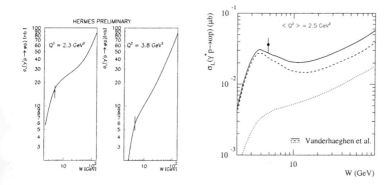

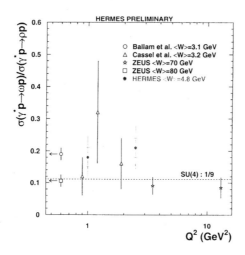

Figure 4. Left: ϕ-production cross section. The curve is the result of a calculation using a 2-gluon exchange mechanism. Right: ω- production cross section. Curves are taken from; short dashes: 2-gluon exchange, long dashes: quark exchange, full line: total.

with the SU(4) predictions for a diffractive mechanism. The HERMES ratio lies above this value and is rather compatible with $2/9$ which would be expected from quark counting rules in the case of a exchange mechanism.

Figure 5. Ratio of the ω to ρ^0 production cross section.

6. t-dependence

At low values of the momentum transfer t to the target nucleon, the cross section falls off exponentially: $\frac{d\sigma}{dt} \propto e^{bt}$. The slope parameter b is known to decrease slowly as Q^2 increases. This reflects the smaller transverse size of

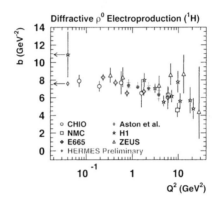

Figure 6.   The Q^2 dependence of the diffractive slope parameter b in exclusive ρ^0 production.

the virtual photon probing the target nucleon, such that at high Q^2 only the size of the target remains to determine the diffractive slope. In fig. 6 the preliminary HERMES data are compared to results from other experiments. The measured Q^2 dependence agrees very nicely with the trend established by the world data. It is interesting to note that the HERMES data are taken at much lower values for W than most of the other data, thus indicating that there is very little W dependence in the slope parameter at finite Q^2. This is in contrast to the situation for real photons where b was seen to increase with increasing W.

References

1. J.Breitweg et al, EPJ C12, 393(2000)
2. C. Adloff *et al*, EPJ C13, 371 (2000)
3. M.Vanderhaeghen, P.A.M.Guichon and M.Guidal, PR D60, 094017 (1999)
4. K.Ackerstaff *et al*, NIM A417, 230 (1998)
5. A.Airapetian *et al*, EPJ C17, 389 (2000)

TRANSVERSE POLARIZATION OF Λ HYPERONS PRODUCED INCLUSIVELY IN EN SCATTERING AT HERMES

O.G. GREBENYUK

Petersburg Nuclear Physics Institute
Gatchina, Leningrad District, 188350, Russia

On behalf of HERMES collaboration

Abstract. The HERMES experiment has measured the transverse polarization of Λ hyperons produced inclusively in quasi-real photon-nucleon scattering at a positron beam energy of 27.6 GeV. The average transverse polarization was found to be $P_n^\Lambda = 0.058 \pm 0.005\,(\text{stat}) \pm 0.006\,(\text{syst})$. The dependence of P_n^Λ on transverse momentum p_T and on the hyperons' light-cone momentum fraction ζ has been investigated. The measured polarization for Λ exhibit different behavior in the current and target fragmentation regions.

1. Introduction

It had been generally believed at the beginning of the Fermilab hyperon program [1] that spin effects in hadronic reactions should be of little importance at high energies. The hard scattering processes, when calculated in perturbative QCD, do not give rise to large polarization effects. It was therefore surprising that a significant negative polarization transverse to the production plane was measured for high energy Λ-s produced by 300 and 400 GeV protons in early Fermilab experiments, the normal to the production plane being directed along $\vec{p}_b \times \vec{p}_\Lambda$ [2, 3]. The absolute value of the polarization increases linearly with the transverse momentum p_t up to the $p_t \simeq 1$ GeV/c.

Presently Λ polarization has been investigated in scattering experiments with a wide variety of beams[4]. Both positive and negative transverse Λ polarization have been observed depending on the beam particle and the

E. Steffens and R. Shanidze (eds.), Spin Structure of the Nucleon, 173–181.

kinematics. In particular the Λ polarization in the K^-p interaction was found to be large and positive.

In terms of the simple quark model, the spin of the Λ corresponds to the spin of the strange quark. Therefore, the presence of the polarization implies that the produced strange quark is polarized. Possible mechanisms for the origin of this polarization are reviewed by Panagiotou in [4]. However there is no common accepted rigorous model to interpret the observed polarizations. At small p_t the quark recombination picture [5] is known to reproduce the experimental tendencies in the inclusive hadron reactions. The model based on quark recombination picture and proposed by De-Grand and Miettinen [6] can account for the relative signs and magnitudes of the polarizations in numerous hadron-to-hyperon transitions. In case of $p \rightarrow \Lambda$ transition the fast valence $(ud)_0$ diquark commonly possessed by projectile and the outgoing Λ recombines with a sea s quark to form the Λ. In case of $K^- \rightarrow \Lambda$ transition the fast valence s quark recombines with sea $(ud)_0$ diquark. The empirical rule conjectured in [6] reads: slow(sea) partons preferentially recombine to fast(valence) partons with their spin down in the scattering plane, whereas fast(valence) partons recombine to slow (sea) partons with spin up. It explains the relative different polarization signs in $p \rightarrow \vec{\Lambda}$ and $K^- \rightarrow \vec{\Lambda}$ transitions. In paper [7] the empirical rule of DeGrand and Miettinen is reproduced with a model relativistic description for the parton-parton interaction realizing the recombination, the polarization being included as the phenomenological parameter.

So far the transverse polarization of Λ's produced by hadrons has been discussed. However it is natural to expect non-vanishing polarization even in the Λ photo-production. It is well known that the real photon has a hadronic component. The quark distributions of the real photon are observed by the deep inelastic lepton-photon scattering (see [8] for reference to the data) and are determined in leading and high order [8]. Hence it is possible to apply the quark recombination approach to the transition $\gamma \rightarrow \Lambda$ [9]. In this case the fast u (or d or s) quark from the photon picks up a a slow (sea) diquark to form the Λ. Both spin-0 and spin-1 diquark recombination processes contribute to this reaction.

The inclusive photo-production of neutral strange particles with the measurement of the transverse Λ and $\bar{\Lambda}$ polarization has in the past been studied at CERN [10] and SLAC [11]. However, the statistics of the published data on Λ lepto- and photo-production is rather poor. The CERN measurements for the incident tagged photon energies between 25 and 70 GeV resulted in an average transverse polarization of 0.06±0.04 for Λ and 0.05±0.10 for $\bar{\Lambda}$. At SLAC the overall polarizations were observed to be 0.09±0.07 for Λ and -0.04±0.04 for $\bar{\Lambda}$ produced with 20 GeV photon beam.

The HERMES experiment offers an excellent opportunity to measure the transverse polarization of inclusively produced Λ's in reaction $eN \to \vec{\Lambda}X$, using the 27.6 GeV positron beam of HERA collider and the internal gas target with a storage cell [12]. Since quasi-real photo-production dominates the inclusive Λ-production at HERMES, it's results are comparable to those of the above mentioned experiments at CERN [10] and SLAC [11], albeit with significantly improved statistics. However the price to be paid for this statististics gain is the absence of the information on the intermediate quasi-real photons since there is no possibility so far to detect in HERMES the scattered positrons with small Q^2. It complicates the interpretation of the data.

HERMES combines both longitudinally polarized and unpolarized targets with the longitudinally polarized electron/positron beam of the HERA storage ring. Due to parity conservation longitudinal beam and target polarizations, P_b and P_t, could give rise to only the double spin asymmetry ϵ in Λ transverse polarization : $P^\Lambda = P_0^\Lambda(1 + \epsilon P_b P_t)$, where P_0^Λ is the Λ transverse polarization measured with unpolarized beam and target. The goal of this experiment was to measure P_0^Λ and, in order to increase statistics, all data of the 1996-2000 data taking period have been included in the analysis, regardless of the longitudinal beam/target polarization. Because of the frequently flipping target helicity in runs with polarized target the influence of the ϵ on P^Λ thus obtained is negligible.

2. Extraction of the transverse Λ polarization and event selection

Parity violation in the weak $\Lambda \to p\pi^-$ decay results in an asymmetry in the angular distribution of the decay proton:

$$\frac{dN}{d\Omega} = \frac{1}{4\pi}\left[1 + \alpha \cos\theta_p P^\Lambda\right], \tag{1}$$

where θ_p is the angle between the proton momentum and the the polarization vector $\vec{P}^\Lambda$ of the Λ in the hyperon rest frame and $\alpha = 0.642 \pm 0.013$ [13] is the analysing power of the parity-violating weak decay.

If neither beam nor target are transversely polarized, the only allowed direction of the Λ polarization is along the normal to the scattering plane formed by the cross-product of the positron momentum, $\vec{p_e}$, and the momentum of the Λ, $\vec{p_\Lambda}$:

$$\hat{n} = \frac{\vec{p_e} \times \vec{p_\Lambda}}{|\vec{p_e} \times \vec{p_\Lambda}|}. \tag{2}$$

A kinematic diagram of inclusive Λ production and decay is given in Fig. 1. The Λ decay is shown in the Λ at rest frame. Although $\hat{n}$ is defined in

O.G. GREBENYUK

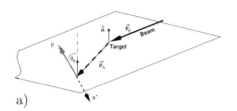

Figure 1. Schematic diagram of inclusive Λ production and decay. The angle θ_p of the decay proton with respect to the normal $\hat{n}$ to the scattering plane is defined in the Λ rest frame.

Eq. 2 using vectors in the laboratory frame, it is important to note that the direction $\hat{n}$ is unaffected by the boost into the Λ rest frame.

For a 4π acceptance the Λ polarization follows from Eq. 1:

$$\{cos\theta_p\} = \alpha\langle cos^2\theta_p\rangle P_n^\Lambda = \frac{\alpha}{3}P_n^\Lambda \tag{3}$$

where $\{...\}$ indicates an average over the experimental data set, while $\langle ...\rangle$ is the corresponding average for an unpolarized Λ sample. In case of restricted acceptance, i.e. when not all protons from Λ decay can be detected in coincidence with accompanying pion, the Eq. 3 transforms to

$$\{\cos\theta_p\} = \frac{\langle\cos\theta_p\rangle + \alpha\langle\cos\theta_p^2\rangle P_n^\Lambda}{1 + \alpha P_n^\Lambda\langle\cos\theta_p\rangle} \approx \langle\cos\theta_p\rangle + \alpha\langle\cos\theta_p^2\rangle P_n^\Lambda, \tag{4}$$

where $\langle\cdots\rangle$ implies an average over an experiment set with unpolarized Λ, where $\langle\cos\theta_p\rangle \neq 0$ is caused by the detector acceptance. Thus, in order to extract P_n^Λ using Eq. 4 one needs to exclude the $\langle\cos\theta_p\rangle$ term.

It can be shown that

$$\cos\theta_p = \frac{-p_{px}p_{\Lambda y} + p_{py}p_{\Lambda x}}{p_{cm}\sqrt{p_{\Lambda x}^2 + p_{\Lambda y}^2}}, \tag{5}$$

where p_{px}, p_{py} and $p_{\Lambda x}$, $p_{\Lambda y}$ are the x- and y-components of the proton and Λ momenta in the laboratory frame. The p_{cm} is the proton momentum in the Λ rest frame.

It is seen from Eq. 5, that $\cos\theta_p$ for two Λ events with $(p_{\Lambda x}, p_{\Lambda y}; p_{px}, p_{py})$ and $(p_{\Lambda x}, -p_{\Lambda y}; p_{px}, -p_{py})$ have the opposite signs. For the up/down mirror symmetric detector the probability to detect both Λ events are the same, which means that

$$\langle\cos\theta_p\rangle^{up} = -\langle\cos\theta_p\rangle^{down} \rightarrow \langle\cos\theta_p\rangle = 0, \tag{6}$$

where the specification up/down can be realized either by the proton ($p_{py} > 0$ - up , $p_{py} < 0$ - down), or by the Λ ($p_{\Lambda y} > 0$ - up , $p_{\Lambda y} < 0$ - down)

For such up/down mirror symmetric detector the Eq. 4 stems

$$P_n^\Lambda \approx \frac{\{\cos\theta_p\}}{\alpha\langle\cos\theta_p^2\rangle}. \tag{7}$$

The $\langle\cos\theta_p\rangle$ term can be excluded also for partially up/down mirror symmetric detector, when the probability to detect $(p_{\Lambda x}, p_{\Lambda y}; p_{px}, p_{py})$ and $(p_{\Lambda x}, -p_{\Lambda y}; p_{px}, -p_{py})$ Λ events differ only by a constant scale factor. Then P_n^Λ can be evaluated as follows

$$P_n^\Lambda \approx \frac{\{\cos\theta_p\}^{up} + \{\cos\theta_p\}^{down}}{2\alpha\langle\cos\theta_p^2\rangle}. \tag{8}$$

The data taken in 1996-2000 years have been used for this study. Since in an inclusive Λ analysis the scattered electron is not detected, direct information on the primary electron vertex is not available. Λ events were identified through the reconstruction of secondary vertices in events with oppositely charged hadron pairs. The minimal distance between the hadron tracks that are combined to a decay vertex $\vec{v}_2 = (x_2, y_2, z_2)$ was required to be less than 1.5 cm. The hyperon track was reconstructed by obtaining the total momentum of the decayed particle and placing the vertex $\vec{v}_2$ in the middle of the segment that provides the minimum distance between the two tracks. In the target cell region the interaction point $\vec{v}_1 = (x_1, y_1, z_1)$ is defined as the middle of the segment that minimizes the distance between the decaying particle track and the beam line. Only events with interaction point inside the target were selected: -22 cm $< z_1 <$ 22 cm and $\sqrt{x_1^2 + y_1^2} < 0.75$ cm. For the data from 1996-1997 years the signal from the threshold Cerenkov detector[12] for the leading hadron was used to separate protons/anti-protons and pions and thus to separate Λ and K_s events. The substantial background still left after the Cerenkov cut was additionally suppressed by the cut $z_2 - z_1 > 10$ cm separating the Λ vertex from the interaction point. Since the 1998 year the Ring Image Cerenkov detector (RICH) is operating at HERMES, the use of which, when analysing the data from 1999 to 2000 years, has provided sufficient background suppression without this vertex separation cut.

3. Results

In this section we present the main results of our analysis. The $p\pi^-$ invariant mass distribution is shown in Fig. 2 after application of the requirements described above. For the polarization analysis, Λ events within a $\pm3\sigma$ invariant mass window of the fitted peak were chosen, and a background-subtraction procedure was applied (as described below). The final data sample contained around 4.17×10^5 Λ events.

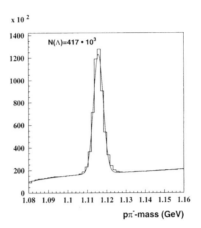

Figure 2.   Invariant mass distributions for Λ.

The contribution to the first and second moments of the $\cos\theta_p$ distribution resulting from the background under the Λ peak has been estimated by a standard side band subtraction.

The transverse polarization for the Λ data samples has been extracted using the formalism of the preceding section (Eq. 8). The final result for the transverse Λ polarization is:

$$P_n^\Lambda = 0.058 \pm 0.005\,(\text{stat}) \pm 0.006\,(\text{syst}).$$

In order to estimate the systematic uncertainty of the measurement, and as a general cross check, identical calculations were carried out for reconstructed h^+h^- hadron-hadron pairs, with leading protons. The events within two mass windows $1.08 < m_{h^+h^-} < 1.11$ and $1.121 < m_{h^+h^-} < 1.2$ GeV (the Λ events being thus excluded) and their secondary vertices placed in the beam-target interaction region have been selected. False polarization values of 0.006 ± 0.001 was found. This value has been used as an estimate of the systematic error on the Λ polarization. As an additional test of the detector's up/down symmetry the K_s sample has been used. The false polarization of K_s was a found to be $P_{K_s} = 0.010 \pm 0.004$.

The good statistical quality of the full inclusive data set allows the dependence of the Λ polarization on certain kinematic variables to be studied. As mentioned earlier, information on the virtual photon kinematics is not available on an event-by-event basis; consequently, only kinematic variables related to the eN system are available. However, one may analyse the data using the kinematic variable $\zeta \equiv (E_\Lambda + p_{z\Lambda})/(E_B + p_B)$, where $E_\Lambda, p_{z\Lambda}$ are the energy and z component of the Λ momentum, and E_B, p_B are the energy and momentum of the beam. This variable is the light-cone momentum fraction of the beam positron carried by the outgoing Λ hyperon. The Λ polarization is shown as a function of ζ in Fig. 3.

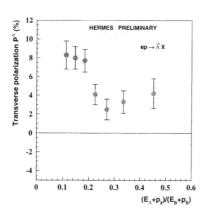

Figure 3.   Transverse polarizations P_n^Λ as a function of $\zeta = (E_\Lambda + p_{z\Lambda})/(E_B + p_B)$.

The dependence of P_n^Λ on the transverse momentum p_T of the Λ has also been explored. Inevitably p_T is defined here with respect to the eN system rather than the $\gamma^* N$ system. It has been noticed that the p_T dependence of P_n^Λ is different for low and high energy Λ's. To separate these two kinematical regimes we suggest to use the kinematical variable $\cos\Theta_\Lambda$, where Θ_Λ is the Λ production angle in the ep CM frame. The idea is that $< p_t > (\cos\Theta_\Lambda)$ has the maximum at $\cos\Theta_\Lambda \simeq 0.6$. Hence taking this cosine as the "watershed" between the low and high energy Λ's, one avoids, at least at small p_T, the combining in a given p_T bin of the low and high energy Λ's. As shown in Fig. 4, a simulation of the reaction using the PYTHIA program reveals that all events at $\cos\Theta_\Lambda \geq 0.6$ are produced in the forward hemisphere $x_F = p_\parallel^\Lambda / p_{max}^\Lambda > 0$.

In Fig. 5 the transverse Λ polarization is shown versus p_T for the two regions: $\cos\Theta_\Lambda < 0.6$ and $\cos\Theta_\Lambda > 0.6$.

4. Discussion

The positive value of the transverse Λ polarization measured by HERMES is surprising, in light of the negative values observed in almost all other reactions. Very few theoretical models of the functional dependence of Λ polarization in photo- or electro-production are available for comparison with the data. One set of curves for the photo-production case is provided in Ref. [9], based on the Quark-Recombination Model. As only the dependence on x_F in the γN frame is calculated in this work, no direct comparison with

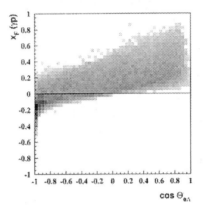

Figure 4. Correlation between x_F, evaluated in the $\gamma^* N$ center-of-mass frame, and the $\cos \Theta_\Lambda$ determined in the eN frame, as determined from a PYTHIA Monte Carlo simulation.

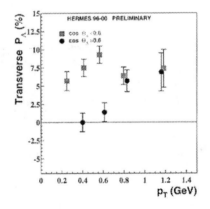

Figure 5. Transverse polarization of Λ hyperons as a function of p_T for two regions: $\cos \Theta_\Lambda < 0.6$ and $\cos \Theta_\Lambda > 0.6$

the present data is possible. However the model does predict a negative polarization in the $x_F^{\gamma N} > 0$ regime, in contradiction with the positive value measured at HERMES for $\cos \Theta > 0.6$. Predictions for the electro-production case are also given in Ref. [14]. Here, transverse Λ polarization is associated with the T-odd fragmentation function $D_{1T}^\perp(z, Q^2)$, one of eight fragmentation functions identified in a complete tree-level analysis of semi-inclusive deep-inelastic scattering [15]. However, these latter calculations are confined to the high-Q^2 regime of deep-inelastic scattering.

One may choose to speculate on the reason for the positive Λ polarization in $\gamma^* N \to \Lambda X$. In the model of DeGrand and Miettinen [6], for example, forward-going Λ's produced in proton-proton scattering are formed from the recombination of a high-momentum valence $(ud)_0$ diquark from the beam with a strange sea quark from the target. The Λ polarization then arises from the acceleration of the strange quark, via the Thomas precession effect. The positive Λ polarization observed with K^- beams is conversely indicative of the deceleration of valence strange quarks from the beam. The positive polarization observed in the HERMES photo-production data might therefore indicate that the $\gamma \to s\bar{s}$ hadronic component of the beam is the dominant source of the polarization of inclusively produced Λ hyperons.

The above discussion concerns the Λ hyperons coming directly from the string. However the contribution from the resonances Σ^0, Ξ and Σ^* should be subtracted. PYTHIA tells that 14% of inclusive Λ's accepted by HERMES are coming from $\Sigma^0 \to \Lambda\gamma$ decay, 29% - from $\Sigma^* \to \Lambda\pi$ decay, the contribution of Ξ being small. Although this simulation needs serious tuning one can take this number as a base. However for the subtraction of the resonances contribution one must know the polarizations of Λ hyperons coming from the dominating resonances and this experiment is in progress at HERMES.

References

1. J. Lach, Nucl. Phys. (Proc. Suppl.) **B50**, 216 (1996).
2. G. Bunce et al., Phys. Rev. Lett **36**, 1113 (1976).
3. K.Heller et al., Phys. Rev. Lett **41**, 607 (1978).
4. A.D. Panagiotou, Int. J. Mod. Phys. **A5**, 1197 (1990).
5. K.P. Das and R.C. Hwa, Phys. Lett. B **68**, 459 (1977),
   R.C. Hwa, Phys. Rev. D **22**, 1593 (1980).
6. T.A. DeGrand and H.I. Miettinen, Phys. Rev. D **23**, 1227 (1981); Phys. Rev. D **24**, 2419 (1981); Phys. Rev. D **32**, 2445 (1985).
7. K.-I. Kubo, Y.Yamamoto and H.Toki, Prog. Theor. Phys. **98**, 95 (1997); H.Toki, N.Nakajima, K.Suzuki and K.-I.Kubo, hep-ph/9906451 (1999).
8. M.G.Glück, E.Reya and A.Vogt, Phys. Rev. D **46**, 1973 (1992).
9. N.Nakajima, K.Suzuki, H.Toki, and K.-I.Kubo, hep-ph/9906451 (1999).
10. D.Aston et al., Nucl. Phys. **B195**, 189 (1982).
11. K.Abe et al., Phys. Rev. D **29**, 1877 (1984).
12. K.Ackerstaff et al., Nucl. Instrum. Methods A **417**, 230 (1998).
13. Particle Data Group, D.E. Groom et al., Eur. Phys. J. C **15**, 1 (2000).
14. M. Anselmino, D. Boer, U. D'Alesio, and F. Murgia, hep-ph/0109186.
15. P.J. Mulders and R.D. Tangerman, Nucl. Phys. **B461**, 197 (1996).

THE COMPASS EXPERIMENT AT CERN

Status and Prospects

G.K. MALLOT [1]

CERN

1211 Geneva 23, Switzerland

1. Introduction

The COMPASS Collaboration brings together about 220 physicists from 27 institutes from Europe, Russia, Japan and India. The goal is to study the structure and spectroscopy of hadrons by a series of experiments with muon and hadron beams in the energy range of 100–200 GeV. The experiment [1] is located at the M2 beam line at the CERN SPS. In 1995 the project started out with two letters of intent which were merged into a common proposal and which was fully approved in September 1998. The first construction phase ended with a commissioning run in 2001. In this article the results of this commissioning run are presented and an outlook on the 2002 run is given.

In the context of this workshop we will focus on the spin programme of COMPASS using a polarised muon beam and a polarised target. The main aim is the determination of the gluon polarisation $\Delta g/g$. In parallel measurements of the flavour-separated Parton Distribution Functions, Δq_f, of the spin transfer in fragmentation ΔD_q^Λ and of $g_1(x, Q^2)$ will be pursued. A determination of the transversity distribution, h_1, is part of the programme with transverse target polarisation.

The gluon polarisation $\Delta g/g$ will be studied using the photon-gluon fusion process $(g\gamma \to q\bar{q})$ shown in Fig. 1. This process will be tagged by either a charmed hadron or a hadron pair with high transverse momentum. The former process is very clean, however difficult to detect, while the latter is abundant but plagued by background from the QCD-Compton process which makes the interpretation to some extent model dependent.

---

[1]On behalf of the COMPASS Collaboration

E. Steffens and R. Shanidze (eds.), Spin Structure of the Nucleon, 183–190.

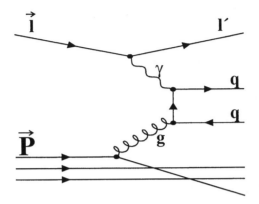

Figure 1.   The photon-gluon fusion process

The charmed quarks are identified by their fragmentation into D^0 mesons and subsequent decay $D^0 \rightarrow \pi^+ K^-$. Other channels like $D^\star$ and three-body decays will also be studied.

2.   The Compass Spectrometer

2.1. OVERVIEW

The COMPASS spectrometer shown in Fig. 2 is an about 60 m long two-stage set-up. Muons of 160 GeV/c momentum and 80 % average polarisation impinge on a double-cell solid-state polarised target (left in Fig. 2). $2 \cdot 10^8$ muons are delivered during the 4.8 s long spill with a repetition time of 16.8 s. Their momenta are measured individually and their trajectory is determined by scintillating fibre hodoscopes and silicon strip detectors upstream of the target. The first stage, the large angle spectrometer, is arranged around a first spectrometer magnet, SM1. Particles emerging from the target are tracked by gaseous detectors, MircoMegas and GEM detectors for the inner region close to the beam and by drift chambers and straw trackers for the outer region. The inner most region is again covered by scintillating fibre hodoscopes. The first spectrometer ends with particle identification detectors: the RICH, the hadronic calorimeter HCAL1 and the muon identification wall MW1.

The second stage, the small angle spectrometer, is arranged around the SM2 spectrometer magnet and tracking is performed by multi-wire proportional chambers and GEM detectors. Again a hadron calorimeter (HCAL2) and a muon identification region (MW2) complete the setup. Trigger hodoscopes and a subsequent logics select the interesting phase space of the scattered muon. Calorimeter information is added for a certain class of physics triggers.

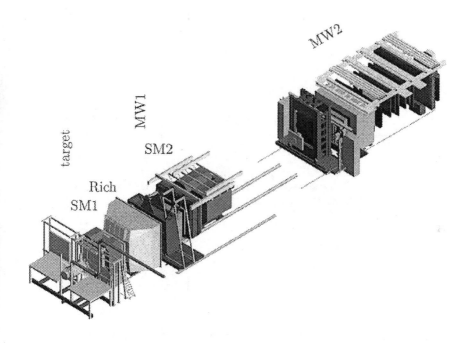

Figure 2.   The COMPASS spectrometer

2.2. THE POLARISED TARGET SYSTEM

Apart from the target material the polarised target is that used by the
Spin Muon Collaboration [2]. In Fig. 3 the muon beam enters from the
left into the two 60 cm long target cells, in which the target material is
polarised oppositely by dynamic nuclear polarisation induced by microwave
irradiation. COMPASS uses ^{6}LiD as deuteron target. The ^{6}Li nucleus can
be modeled as a ^{4}He core plus a deuteron, leading to a very favourable
fraction of polarisable nucleons of $f = 0.5$. A maximum polarisation of
55 % was achieved in the 2001 run, however average polarisation in the
target was rather around 48 %. The polarised target system works with
a magnetic field of 2.5 T. The spin orientations are inverted every 8 h by
a rotation of the magnetic field vector using a superposition of solenoid
and dipole fields. The ^{3}He/^{4}He dilution refrigerator reaches temperatures
of 50 mK in frozen spin mode. In transverse mode the polarisation is held
by a 0.5 T dipole field.

COMPASS plans to use a new magnet with an aperture of ±180 mrad.

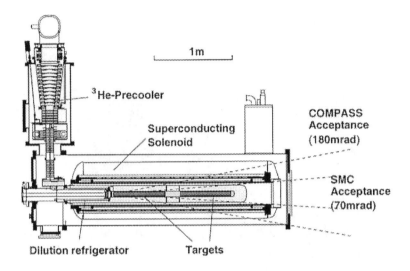

Figure 3.   The polarised target

The completion of this magnet is delayed. The currently used SMC magnet provides a limited acceptance of ±70 mrad. However, at the presently adopted muon energy of 160 GeV this leads only to a 20 % lower event yield for the D^0 production. For other channels, like transversity, the restrictions are more severe.

3.   Selected spectrometer elements

3.1.   NOVEL GASEOUS TRACKERS

Two types of new tracking devices are used in the COMPASS experiment, the *MICRO MEsh GAseous Structure* (Micromegas) [3] and the *Gaseous Electron Multiplier* (GEM) [4]. Both are based on a fast collection of the slow positive ions, which gives high rate capabilities and thus allows their operation in the intensive beam region. The principle of the detectors is depicted in Figs. 4 and 5. In the Micromegas the ionisation or conversion region is separated from the amplification region by a metallic mesh. While the former has a typical thickness of a few mm the latter is constraint to typically 100 μm yielding electrical field strength of 40 kV/cm. In a GEM the amplification takes place within the 50 μm thickness of a copper-coated caption foil. The electron avalanche is created in 70 μm diameter holes of a perforated kapton foil where the electrical field strength is about the same as for the Micromegas. In COMPASS triple GEM arrangements are used to reduce the amplification in a simple GEM. The read-out PCB

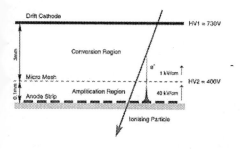

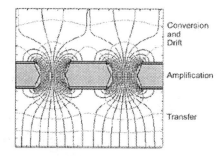

Figure 4.   Schematics of the MicroMegas detector

Figure 5.   Schematics of the GEM

has crossed strips for two-dimensional read-out. The x and y-coordinate can be correlated by the signal amplitude, which is measured using ADCs developped for silicon detector read-out. The active area of the detectors are $40 \cdot 40$ cm^2 and $32 \cdot 32$ cm^2 for the Micromegas and GEMs, respectively. Both have a inner dead zone of 5 cm diameter where the muon beam passes. For both the spacial resolution is in range of 50–70 μm and the efficiency about 97 %.

3.2. THE RICH DETECTOR

The COMPASS RICH-1 [5] is situated in the large-angle part of the spectrometer and is designed to separate kaons and pions up to about 60 GeV/c. It uses C$_4$F$_{10}$ gas as radiator. The Cherenkov light is reflected by an upper and a lower spherical mirror surface comprising 116 VUV reflective hexagonal mirrors and covering 21 m^2. The photon detectors are based on MWPCs with CsI photocathodes operating in the wavelength range of 160–200 nm. The total photodetector surface is 5.3 m^2 divided into 84000 analog read-out pixels. While in 2001 only 50 % of the radiator gas was available, the RICH could entirely be filled during the 2002 run. In Fig. 6 the preliminary particle identification data from Rich-1 are shown for a low intensity physics run. The data are dominated by the huge pion ridge, but also the kaons are well visible. Electrons show up at large angles and low momenta.

3.3. THE DATA ACQUISITION

The COMPASS DAQ is based on a pipelined readout architecture. The data are digitized on the front-end cards located on the detectors and are subsequently buffered in about 160 CATCH modules. From there data are transferred via S-links to 16 read-out buffer computers (ROBs). The actual

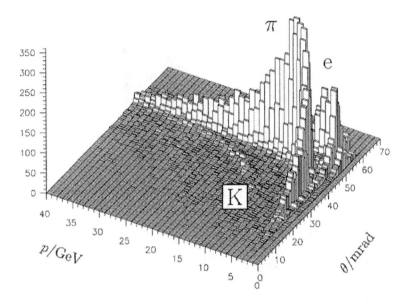

Figure 6.   Preliminary RICH data: particle momentum vs measured Cherenkov angle.

event building takes place on 12 event builder computers connected to the ROBs via Gigabit Ethernet and powerful network switchs. From there the data are sent to the central data recording in the CERN computer centre. The DAQ deals with about 250 000 read-out channels and routinely handled a trigger rate of 5 kHz during the 2002 run. Tests were performed up to a rate of 50 kHz. The typical event size is 40 kByte yielding data rates of 1 GByte per SPS spill or 3–4 TByte/day.

4.   The 2001 and 2002 runs

The commissioning run in 2001 consisted of a set-up period from July to October and a final two week period of data taking with all available detectors. However, only about 30 % of the detectors of the full initial layout were installed and redunancy was low. Nevertheless, the new detectors and the DAQ system were fully commissioned and the alignment and reconstruction procedures tested successfully.

For the 2002 run the complete set of detectors (apart from a few straw planes) were installed and the full redundancy of the spectrometer was available. After a tuning phase full data taking proceeded from July to mid September. In total about $5 \cdot 10^9$ events were recorded. The total data volume is 260 TByte. The reconstruction of the huge data sample poses a

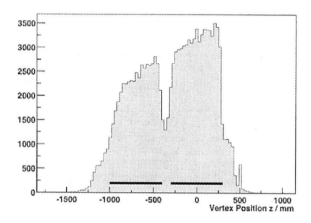

Figure 7.   Primary interaction vertices in the two target cells.

formidable challenge.

In Fig. 7 the distribution of the primary vertices is shown along the beam direction. The two oppositely polarised target cells are also indicated. First preliminary results were obtained on the production of ρ, K^0, Λ, and ϕ, where the latter involves kaon identification by the RICH. The Armenterosa-Podolanski plot (Fig. 8) shows the asymmetry of the longitudinal momenta of positive and negative particles emerging from a secondary vertex versus their transverse momenta. The phase space corresponding to the decay of neutral particle is an ellipse in this plot. Clearly the symmetric K^0 arc and the two arcs corresponding to $\overline{\Lambda}$ (left) and Λ (right) are visible. The mass resolution is 6 MeV for the K^0 and 2.7 MeV for the Λs, close to the values from the Monte-Carlo simulations.

5. Summary and outlook

The COMPASS experiment has successfully been commissioned in 2001 and a first full period of data taking in 2002 yielded a large data sample. During 2002 the availability of the experiment was high and the spectrometer operated reliably. About 20 % of the data were taken with transverse target polarisation. In 2003 and 2004 COMPASS will run for about 80–100 days per year, depending on the schedule of the CERN SPS. CERN accelerators will be shut down during 2005. CERN has made provision for COMPASS running in the years 2006–2010. However, the final approval has to wait until COMPASS has spelled out the programme for this period in detail.

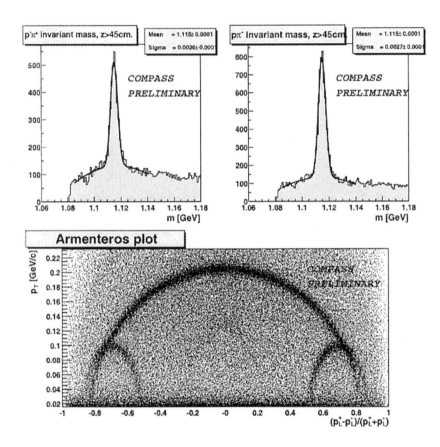

Figure 8.   top: pion-kaon invariant mass distributions with $\overline{\Lambda}$ and Λ peak. bottom: Armenterosa-Podolanski plot, see text

Acknowledgments

I thank the organisers of this inspiring workshop for the opportunity to present our experiment and my COMPASS colleagues M. Leberig and A. Korzenev, for providing me most of the plots.

References

1. G. Baum *et al.*, COMPASS proposal, (1996) *CERN-SPSLC-96-14*
2. D. Adams *et al.*, (1999) *Nucl. Instrum. Meth.*, **A437**, pp. 23–67
3. D. Thers *et al.*, (2001) *Nucl. Instrum. Meth.*, **A469**, pp. 133–146
4. F. Sauli, (1997) *Nucl. Instrum. Meth.*, **A386**, pp. 531
5. G. Baum *et al.*, (1999) *Nucl. Instrum. Meth.*, **A433**, pp. 207

CHARM PRODUCTION AT HERA USING THE ZEUS DETECTOR

G.G. AGHUZUMTSYAN
Physikalisches Institut der Universität Bonn
Nussallee 12, 53115 Bonn

Abstract.

Recent results on the production of $D^{*\pm}(2010)$ mesons in both photo-production (PHP) and deep inelastic scattering (DIS) are reviewed. The decay channel $D^{*+} \to D^0 \pi^+ \to K^- \pi^+ \pi^+$ and its charged conjugate were used to identify the D^* mesons. The results are compared with theoretical predictions.

1. Introduction

The scattering process of a lepton and a proton can be described by the exchange of an electroweak boson: a photon, a Z boson or a W boson. The boson-gluon fusion diagram that dominates charm production is shown in Figure 1.

At HERA, which is an electron - proton collider, the beam energies of 27.5 GeV for the electron and 920 GeV the for proton, imply a centre of mass energy, $\sqrt{s}$, 318 GeV. A full specification of the process requires two variable usually taken as $Q^2 = -q^2 = (k - k')^2$, which is is the negative squared four momentum transfer from the incoming electron to the proton, and Bjørken $x = \frac{Q^2}{2P \cdot q}$, which is the momentum fraction of the proton carried by the struck parton. The fractional energy transfered to the proton in its rest frame is $y = \frac{P \cdot q}{P \cdot k}$, where $Q^2 = sxy$. The

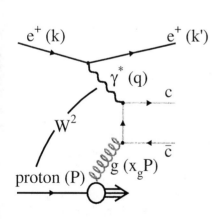

Figure 1. Sketch of the boson-gluon-fusion process in $e^+ p$ collisions

E. Steffens and R. Shanidze (eds.), Spin Structure of the Nucleon, 191–199.

invariant mass of the $\gamma - p$ system is given by $W = (P + q) \cong \sqrt{(4E_e E_p y)}$. Two kinematic regimes are selected. The first is deep inelastic scattering (DIS), when $Q^2 > 1\,\mathrm{GeV}^2$. In this case the scattered electron is visible in the main detector and the exchanged boson is highly virtual. The second is photoproduction (PHP), when $Q^2 \leq 1\,\mathrm{GeV}^2$ with typical values of $10^{-3}\,\mathrm{GeV}^2$. In this case there is no scattered electron in the main detector and the exchanged boson is a quasi-real photon.

Both kinematic regions are interesting for studying charm production and its interpretation in terms of QCD. For perturbative QCD (pQCD) calculations to be valid a hard scale is needed. The scale at which the coupling gets too large in order to apply perturbative calculation techniques, Λ_{QCD} is of the order of 200 MeV [1].

Quarks with a mass which is considerably higher ($m_Q \gg \Lambda$) are regarded as heavy quarks. Since $m_c \sim 1.5\,\mathrm{GeV}$ charm production is a good testing ground for pQCD. In addition to the heavy quark (HQ) production mechanism the proton and photon structures can be studied.

The typical process of charm production, which has a QCD leading order (LO) contribution to open HQ production, is shown in figure 1. In this case whole the photon energy participates in the hard subprocess. couples to a gluon from the proton. The final state of the hard subprocess consists of a $q\bar{q}$ of two partons. Another possibility, common in photoproduction, is resolved charm production where the photon acts as a source of partons. One parton from the photon interacts with a parton from the proton in the hard subprocess. Thus it is similar to hard hadron-hadron interactions. The processes give a direct handle on the gluon density in proton, $x_g(x)$, and the partonic structure of the photon. Charm quarks present in the parton distribution functions (PDF) of the photon and proton lead to processes of the form $cg \to cg$ and $cq \to cq$, which are referred to as "charm excitation".

2. Charm quark production models

Two charm production models have been developed. One of them is based on next-to-leading-order (NLO) QCD calculations of order α_s^2 in the coefficient functions in the so-called fixed-flavour number scheme (FFNS). In this scheme (massive scheme) the number of active quark flavours is fixed, independent of Q^2. Only the three light quarks (u, d, s) and gluons contribute to the photon and proton structure function. There is no explicit c-excitation component. Charm quarks are produced dynamically. The presence of the two large scales, Q^2 and m_c^2, makes this scheme unreliable in the region $Q^2, p_t^2 \gg m_c^2$ because the neglected terms of orders higher than α_s^2 contain $\log(p_t^2/m_c^2)$ factors that become large and spoil the convergence of the perturbation series. The massive scheme is therefore reliable in the region $p_t^2 \sim m_c^2$. The other charm production model is the so-called massless model, where the heavy quarks contribute to the PDF. Above a certain threshold charm is treated as an active flavour with zero mass. This sceme is called the variable-flavour-number scheme (ZM-VFMS). Because of that, in addition to the process $gg \to c\bar{c}$ and $q\bar{q} \to c\bar{c}$, additional charm-excitation leading order processes are present in this scheme: $qc \to qc$, $gc \to gc$. The massless scheme allows the resummation of the terms $\log(p_t^2/m_c^2)$ by providing the perturbative fragmentation function for the heavy quark. This fragmentation function includes the logarithmic dependence. The massless method is not reliable at $p_t^2 \sim m_c^2$ and can only be expected to produce reasonable predictions for $p_t^2 \gg m_c^2$. A comparison of the two methods has been performed [2].

3. Charm identification via D* decay

For the results presented in this paper the D^* decay method was used for the identification of charm quarks. The branching ratio of the charm quark to a $D^{*\pm}$ meson was measured at OPAL [3], ALEPH [4] and DELPHI [5]. According to [6] the average is $f(c \to D^{*+}) = 0.235 \pm 0.007$.

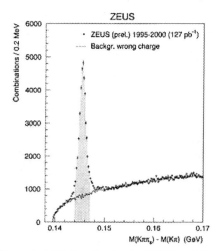

Figure 2. ΔM distribution in PHP for 1995-2000 data

The D^* are identified using a standard ΔM mass difference reconstruction method in two different channels $D^{*\pm} \to D^0\pi_s \to (K^-\pi^+$ or $K^-\pi^+\pi^-\pi^+)\pi_s^+$ and the charge conjugate. Since the branching ratios for the channels are known, it is possible to measure just one channel and extrapolate to the total number of D^* mesons. Figure 2 shows the distribution of ΔM for candidates with $M(K\pi)$ close to $M(D^0)$ for the $K2\pi$ final state in photoproduction in the 1995-2000 data. The signal is clear over a small background. Similar distributions we observed in DIS.

4. Open charm production in DIS

The production of $D^{*\pm}$ mesons in deep inelastic scattering has been measured at ZEUS using an integrated luminosity of 83 pb^{-1}. Figures 3a and 3b show the differential cross section for D* production as a function of $\log Q^2$ and $\log x$. The data are compared with NLO QCD calculations as

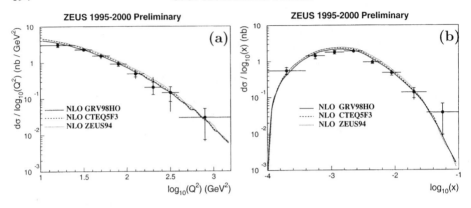

Figure 3. Differential cross sections for $D^{*\pm}$ candidates as a function of Q^2 a) and x b) compared with NLO QCD calculations.

implemented in HVQDIS [7] program using the NLO parton distribution functions (PDF).

In DIS it is also possible to measure the charm contribution $F_2^{c\bar{c}}$ to the proton structure function by extrapolating the cross sections in Q^2 and y bins to the full $p_t(D^*)$ and $\eta(D^*)$ phase space using HVQDIS with the RAPGAP-based fragmentation [8]. Then using the hadronisation fraction of charm to D^{*+} [3] the extrapolated cross sections were converted into charm cross sections.

Figure 4 shows the ratio of $F_2^{c\bar{c}}$ to F_2, the inclusive proton structure function, as a function of x in fixed Q^2 bins. The curves represent the results of NLO QCD calculations with the ZEUS NLO QCD PDFs. The central, solid curve uses 1.5 GeV for the charm mass and the dashed curves correspond to a charm mass variation between $m_c = 1.3$ and $m_c = 1.7$ GeV. The ratio rises with decreasing x; for small Q^2 the contribution of $F_2^{c\bar{c}}$ to F_2 is about 10%, while for high Q^2 at low x it increases to about 30%.

5. Charm photoproduction

Charm photoproduction has been measured both with a wide band beam and with the scattered electron in a dedicated tagger (tagged photoproduction). Figures 5a and 5b show the differential cross section as a function of $p_t(D^*)$ and $\eta(D^*)$ using the 44 meter tagger. This corresponds to the region $80 < W < 120$ GeV. In figure 5(a) the differential cross section is compared with the calculation of Kniehl et al. [9, 10, 11], which was performed with three different photon structure function parametrisations. The continuous line was calculated with GRV-G HO [12], the dashed with GS-G-96 HO and the dotted with AFG-G HO [13]. There is reasonable agreement

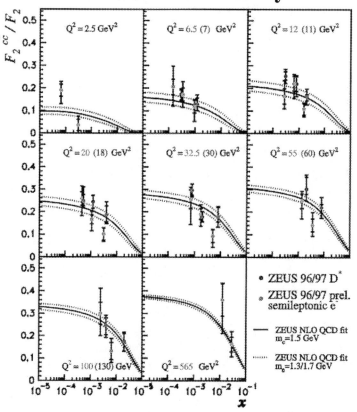

Figure 4.   The charm structure function in neutral current DIS.

between the data and all of the calculations. Figure 5b shows the results
of the massless calculations with different photon structure functions. The
calculations are sensitive to the photon structure function parametrisations
used. Compared with the massive case this massless calculation is about a
factor of two higher and agrees better with the measured data [14]. But in
both cases there is a large discrepancy in the forward (proton) direction,
$\eta^{D^*} > 0$. A similar discrepancy was observed in the ZEUS untagged D^*
analysis [15].

The D^* photoproduction data can also be used to study the charm
content of the photon structure.

The fraction of photon momentum participating in two high E_t jets,
produced in the final state is given by $x_\gamma^{OBS} = \frac{\Sigma_{jets} \cdot E_T \cdot e^{-\eta}}{2y \cdot E_e}$, where $y \cdot E_e$ is
the energy of the exchanged photon and the sum runs over the two highest
E_t jets in the event within the acceptance η. This definition of x_γ^{OBS} is well

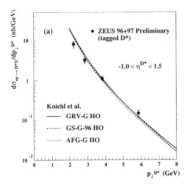
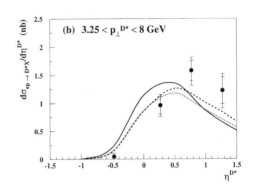

Figure 5. Low-W differential cross sections $d\sigma/dp_\perp^{D^*}$ a) and $d\sigma/d\eta^{D^*}$ b) compared to massless NLO calculations with several photon structure function parametrisations.

defined in all orders of perturbation theory. In the experimental data, in the NLO calculations and in the Monte Carlo distributions for direct and resolved enriched samples are defined by a cut on x_γ^{OBS} with $x_\gamma^{OBS} < 0.75$ defining the resolved sample.

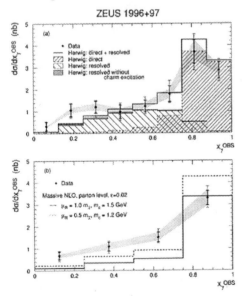

Figure 6. The ZEUS $d\sigma/dx_\gamma^{OBS}$ cross section for D^* mesons

Figure 6a clearly shows a large contribution coming from resolved events. A significant leading order resolved component is required in order to describe the data of which ($\sim$95%) is due to charm excitation in the photon. Figure 6b allows the comparison with massive NLO theory. The calculation lies below the data at low x_γ^{OBS} which, together with the discrepancy in the inclusive cross-section in the forward direction, may be an indication of some deficiency in the current available parametrisations of the photon structure [15].

The large resolved contribution, allows a study of the dynamics of charm production using the angular distributions of charm. For a gluon propagator we expect a steep rise is expected, because the angular dependence of cross sections for resolved processes is proportional to $\sim (1 - \cos\theta^*)^{-2}$ [16], where $\theta^* = \tan(0.5(\eta^{jet1} - \eta^{jet2}))$ is the angle between the jet-jet axis and the beam axis in the dijet rest frame. For a quark

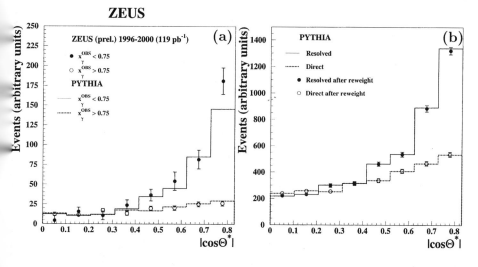

Figure 7. Differential distributions $dN/d|\cos\theta^*|$ in PHP. a) differential distribution of the data and of the PYTHIA MC simulation; b) differential distribution for Monte Carlo with the true LO definition of direct/resolved. Results are given separately for direct photon (open dots/dashed lines) and for resolved photon (black dots/full histogram) events.

propagator the angular distribution varies as $(\sim (1 - \cos\theta^*)^{-1})$. The expectations we born out in Figure 7 which shows the event distribution as a function of $|\cos\theta^*|$. For low x_γ^{OBS} a steeper rise towards high $|\cos\theta^*|$ is evident. That suggests, that the dominant propagator in resolved processes is a gluon propagator as expected, which is the signature for charm excitation. For the direct process the distribution is rises less steep, which is the signature for a quark propagator (quark exchange as predicted by QCD).

6. ZEUS tracking 2000 upgrade

With the planned luminosity upgrade (a factor of five) in addition to the tracking detectors at ZEUS new silicon microvertex detector(MVD) and straw-tube transition radiation detector/tracker (STT) (Figure 8) have been installed at HERA during year 2000 upgrade. The new detectors require upgrades to the tracking, to the pattern recognition and to particle identification. A global tracking trigger (GTT) [17] has also been designed and implemented.

The MVD is designed to improve the tracking resolution and allow the reconstruction of secondary vertices from heavy quark decays with a minimal background. The detector consists of a long "Barrel" with three layers of Si detectors parallel to the beam. In the forward region there are four layers of silicon detectors, perpendicular the beam. The detector is

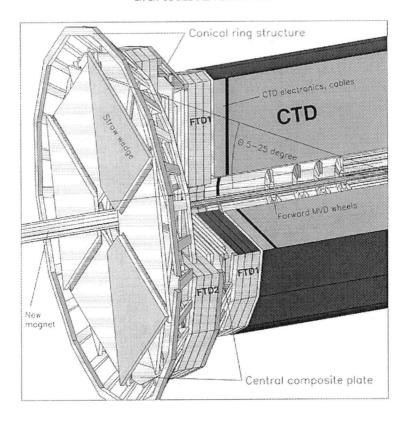

Figure 8.   ZEUS tracking trigger upgrade

designed to be both light and stiff to for achieve high position accuracy
(5-10μm) [18].

The STT, in addition to the existing central tracking detector (CTD)
will significantly extend the coverage of rapidity and improve the tracking
efficiency in the forward region. The detector consists of 8 superlayers. Each
superlayer has 6 sectors and each sector comprises 192(small) or 266(large)
straws. The STT covers the full azimuthal angle and the polar angular
range from 6 to 23 degrees [19].

Figure 8 shows a 3-D view of the central and forward tracking detectors
at ZEUS. The tracking upgrade allows the CTD+MVD+STT detectors to
be combined, which will improve the track reconstruction efficiency and
lied to a substantial improvement in the charm tagging efficiency.

7. Summary

Some charm DIS and photoproduction results from ZEUS has been reviewed. The data are compared with theoretical predictions. Results on

charm production in DIS are reasonably well described by the NLO QCD calculations. Differential cross sections for the low W range in PHP have an excess in the forward(proton) direction. Similar results were found in the ZEUS untagged D^* analysis. The luminosity upgrade will allow us to investigate this discrepancy. Interesting results in dijet angular distribution in photoproduction are reviewed, which gives us the possibility to distinguish between direct and resolved PHP. A large contribution of charm-excitation in the resolved processes is seen, which corresponds to a gluon propagator.

With the luminosity upgrade at HERA in 2000 the new tracking detectors MVD and STT were installed, which were designed to improve the tracking resolution, measure the secondary vertices and extend the coverage of rapidity in the forward direction.

8. Acknowledgements

I would like to thank the local organisers for a very useful and enjoyable workshop.

References

1. R.K. Ellis, W.J. Stirling and B.R. Webber, *QCD and Collider Physics*, Cambridge Monographs on Particle Physics, Nuclear Physics and Cosmology, Vol. 8. Cambridge University Press, 1996.
2. ZEUS Coll., M.Cacciari, *On heavy quarks photoproduction and $c \to D^*$ fragmentation functions.* To appear in the Proceedings of the Ringberg Workshop "New Trends in HERA Physics", 25-30 May 1997, 1997.
3. OPAL Collab., R.Akers et al., Z. Phys. **C 67**, 27 (1995).
4. ZEUS Coll., D.Buskulic et al., Z. Phys. **C 62**, 1 (1994).
5. DELPHI Coll., P.Abreu et al., Z. Phys. **C 59**, 533 (1993).
6. L. Gladilin, Preprint hep-ex/9912064, DESY, 1999.
7. ZEUS Coll., $D^{*\pm}$ *production in DIS.* Abstract 855, XXXth International Conference on High Energy Physics, Osaka, Japan, 2000, 2000.
8. ZEUS Coll., J. Breitweg et al., Eur. Phys. J. **C 12**, 35 (2000).
9. B.Kniehl et al., Z. Phys. **C 76**, 689 (1997).
10. J.Binnewies, B.Kniehl and G.Kramer, Z. Phys. **C 76**, 677 (1997).
11. J.Binnewies, B.Kniehl and G.Kramer, Phys. Rev. **D 58**, 014014 (1998).
12. M. Glück, E. Reya and A. Vogt, Phys. Rev. **D 45**, 3986 (1992).
13. P. Aurenche, J.P. Guillet and M. Fontannaz, Z. Phys. **C 64**, 621 (1994).
14. S. Stonjek, *Measurement of $D^{*\pm}$ Cross Sections in Photon Proton Collisions at HERA.* Ph.D. Thesis, Hamburg University, Hamburg (Germany), Report DESY-THESIS-2001-043, 2001.
15. ZEUS Coll., J. Breitweg et al., Eur. Phys. J. **C 6**, 67 (1999).
16. H. Abramowicz and A. Caldwell, Rev. Mod. Phys. **71**, 1275 (1999).
17. ZEUS Coll., M. Sutton, C. Youngman, *The Global Tracking Trigger System and Architecture* (unpublished). ZEUS-Note-99-074, 1999.
18. ZEUS Coll., *A Microvertex Detector for ZEUS* (unpublished). ZEUS-Note-97-006, 1997.
19. ZEUS Coll., *A Sraw-Tube Tracker for ZEUS* (unpublished). ZEUS-Note-98-046, 1998.

SPIN PHYSICS OF VECTOR MESON PRODUCTION AT HERMES

A.BORISSOV

Randall Laboratory of Physics, University of Michigan,
Ann Arbor, MI 48109-1120, USA.

Abstract. The exclusive diffractive production of light vector mesons (ρ^0 and ϕ) in the intermediate energy region at moderate Q^2 is described. Double-spin asymmetries in the cross section for ρ^0 and ϕ mesons have been measured in electroproduction and quasi-real photoproduction on proton and deuterium targets. Moreover, spin density matrix elements have been determined for exclusive diffractive ρ^0 and ϕ production, which are compared to results obtained at higher energy. From these data the helicity transfer from the virtual photon to the vector mesons, the ratio of the longitudinal to transverse components of the production cross section, and a test of the s-channel helicity conservation have been obtained.

1. Introduction

The HERMES experiment [1] was designed to study the spin structure of the nucleons through semi-inclusive Deep Inelastic Scattering (DIS) measurements using HERA's polarized electron (positron) beam and a polarized internal gas target. For the study of semi-inclusive DIS reactions this spectrometer has a rather large acceptance and as such it is also suitable for the detection of the decay products of vector mesons. This allows to study spin effects in vector meson production and to get a comprehensive picture of such processes in the exclusive limit at intermediate energies.

For that reason, HERMES data have been used for extensive studies of diffractive production of light vector mesons in the intermediate energy region ($4.0 < W < 6.0$ GeV) at moderate momentum transfer squared ($0.7 < Q^2 < 5.0$ GeV^2). In this domain, an interplay between processes studied up to now separately in low energy experiments ($W < 4$ GeV) and high energy experiments ($W > 10$ GeV) occurs. Theoretically, elastic vector meson electroproduction is usually described differently in the two

E. Steffens and R. Shanidze (eds.), Spin Structure of the Nucleon, 201–210.

domains. At low energies, the hadronic component of the photon can be seen as a superposition of the lightest vector mesons, a process, which is well described by the Vector Dominance Model (VDM). At higher energies, models based on perturbative QCD (pQCD) are more appropriate. New data, in particular cross section measurements [2], double spin asymmetries [3, 4], and spin density matrix elements [5, 6] will allow us to obtain more insight in the production of vector mesons at intermediate energy. In this paper such data are presented and compared to high (or low) energy data and to available theoretical calculations.

A detailed description of the HERMES experiment can be found in Ref. [1]. It is sufficient to note here that the experiment consists of a forward angle spectrometer with an angular acceptance ranging from +(-)40 to +(-)140 mrad in the vertical direction, and from -170 to +170 mrad in the horizontal direction. A dipole magnet, various sets of tracking chambers, and various particle identification detectors provide information on the direction, momentum, charge and type of the particles detected. Behind these detectors, a large segmented electromagnetic calorimeter is located, which is used for electron-hadron discrimination and for photon identification.

The longitudinal polarization of the 27.5 GeV HERA electron (positron) beam is obtained by spin rotators located upstream and downstream the HERMES experiment and has values up to about 60%. Both polarized and unpolarized gases can be injected into the HERMES target cell. During the first HERMES running period (1995-2000) measurements were performed on longitudinally polarized proton and deuterium targets with polarization levels of typically 90%. Data collected in the years 1996-1997 on a proton target and first results obtained on a deuterium target in 1998-2000 are presented here.

The paper is organized as follows. New data on ρ^0 and ϕ double spin asymmetries obtained on proton and deuterium targets are presented in section 2. Spin density matrix elements for ρ^0 and ϕ electroproduction are shown in section 3 and compared to data from higher energy experiments. In the last section the common picture emerging from these exclusive data is discussed.

2. Double Spin Asymmetry

At the HERMES experiment both the beam and target are polarized. Hence, a double spin asymmetry for the production of vector mesons can be derived from the data. The cross section asymmetry is defined as $A_1^{\rho^0} = \frac{\sigma_{1/2} - \sigma_{3/2}}{\sigma_{1/2} + \sigma_{3/2}}$, where $\sigma_{1/2}$ and $\sigma_{3/2}$ correspond to the cross sections for the two different spin projections of the virtual photon and the target nucleon on the beam direction. The measured asymmetry is defined as

$A_\parallel^{\rho^0} = \frac{\sigma_{\uparrow\downarrow} - \sigma_{\uparrow\uparrow}}{\sigma_{\uparrow\downarrow} + \sigma_{\uparrow\uparrow}}$, where $\sigma_{\uparrow\uparrow}$ ($\sigma_{\uparrow\downarrow}$) denotes the cross section when the spin orientations of beam and target are aligned anti-parallel (parallel). The relationship $A_\parallel^{\rho^0} = D \cdot (A_1^{\rho^0} + \eta A_2^{\rho^0})$ allows to obtain the double spin asymmetry $A_1^{\rho^0}$ from the data taking into account the photon depolarization factor D and a small contribution due to the transverse asymmetry $\eta A_2^{\rho^0}$ [3]. A positive asymmetry $A_1^{\rho^0} = 0.244 \pm 0.111_{stat} \pm 0.021_{syst}$ was found in exclusive ρ^0 production at HERMES using a polarized ^{1}H target data [3].

This result is compatible with the prediction of Ref. [7]: $A_1^{\rho^0} / A_1^{p(d)} \simeq 2/(1 + (A_1^{p(d)})^2)$, where $A_1^{p(d)}$ is the inclusive asymmetry measured in deep inelastic scattering for proton (deuterium) targets. This relationship has been verified with data for both targets as shown in Fig. 1, top panels. The good agreement for the ratio of $A_1^{\rho^0}$ and $A_1^{p(d)}$ on both targets suggests the presence of production amplitudes corresponding to unnatural parity exchange in the t-channel. Unnatural parity exchange might originate from diquark-exchange which would agree with the previously made conclusion that ρ^0 production at HERMES is dominated by quark-exchange [2]. At higher W values, the SMC collaboration finds an asymmetry compatible with zero [8]. As the inclusive asymmetry in their kinematic region is much smaller, one would expect a lower asymmetry based on the previously cited model. Furthermore, at SMC kinematics, the production cross section is dominated by gluon-exchange, so the difference in reaction mechanisms may well be reflected in the spin asymmetry.

The double spin asymmetry has also been determined in the so-called quasi-real photoproduction regime. The selection criterion were two accepted in the spectrometer tracks, corresponding to the products of vector meson decay. The scattered positron could not detected as it passes through the beam pipe. The average Q^2 was about 0.1 GeV^2, as concluded from Monte Carlo simulations, and no separation between elastic and inelastic production of ρ^0 mesons was performed. On the proton target, the double spin asymmetry for quasi-real photoproduction of ρ^0 mesons is $A_1^{\rho^0} = 0.015 \pm 0.017$, and on the deuterium target $A_1^{\rho^0} = -0.004 \pm 0.006$, i.e. the asymmetries are found to be consistent with zero. The asymmetry for quasi-real photoproduction of ϕ mesons was also found to be close to zero, i.e. $A_1^{\phi} = 0.07 \pm 0.11$ on the proton, and $A_1^{\phi} = 0.02 \pm 0.04$ on the deuterium target. Note that from the small statistics presently available for ϕ electroproduction on polarized targets, no significant double spin asymmetry could be extracted.

Systematic studies included the comparison of the data to Monte Carlo simulations, different methods for background subtraction, and studies of the stability of the extracted spin asymmetry with time. The systematic

A.BORISSOV

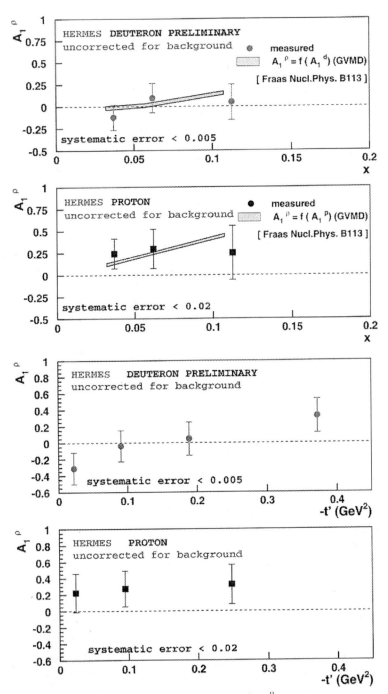

Figure 1. In the top panel the double spin asymmetry $A_1^{\rho^0}$ on deuterium (top part) and proton (bottom part) targets as a function of x_{Bj} are compared to theoretical calculations [7]. On the bottom panel the t'-dependence of $A_1^{\rho^0}$ is presented for deuterium (top part) and proton (bottom part) targets.

uncertainty (of typically 0.002) is mainly due to the target polarization measurement and the presently unknown transverse asymmetry for ρ^0 production. The kinematical dependence of the double spin asymmetry $A_1^{\rho^0}$ was investigated for four different variables covering the following range: $0.5 < Q^2 < 5$, GeV^2, $4 < W < 6.0$ GeV, $0.02 < x_{Bj} < 0.2$, and $0 < -t' < 0.4$ GeV^2. No significant dependence was observed in any of the first three variables when using the proton target [3]. Data on the deuterium target suggest a possible t'-dependence of the asymmetry as displayed in Fig. 1, bottom panels.

3. Spin Density Matrix Elements

Angular distributions of the products of vector meson decays are influenced by the transfer of the virtual photon's spin to that of the vector meson produced. Previous experiments have shown that this is most conveniently presented in the s-channel helicity frame, i.e. the center-of-mass-system of the $\gamma^* p$ system, where the quantization axis is chosen as being opposite to the direction of the recoiling target in vector meson rest frame. Only in this reference frame matrix elements are not strongly t-dependent. The decay angle distribution can be expressed in terms of the polar and azimuthal angles Θ and ϕ relative to the quantization axis, with the azimuth defined relative to the vector meson production plane. The third angle necessary to fully characterize the angular distribution is the angle Φ between the electron scattering plane and the vector meson production plane. The full expression for the decay angular distributions can be found in Ref. [9], where it is given in terms of the polarization density matrix $\rho_{\lambda\lambda'}$ of the vector meson. Here only expressions for angular distributions integrated over two of the three angles are used, which correspond to linear combinations $r_{\lambda\lambda'}^\alpha$ of the spin density matrix elements.

Some results for the spin density matrix elements from ϕ production are shown in Figs. 2 and 3 [5]. The data on ϕ production appear to exhibit s-channel helicity conservation, see Fig. 2. Note that the limited statistics obtained for ϕ production restricts the possibility of comparing measured spin density matrix elements to those obtained in other experiments. For example, the measured spin density matrix elements for ϕ meson production at HERMES are consistent with H1 and ZEUS data [10].

The extracted spin density matrix elements from the (decay) angular distributions for ρ^0 production are displayed in Fig. 4 [6]. In most cases the data are compared to a solid vertical line, which represents the value of that matrix element if s-channel helicity conservation is assumed. The s-channel helicity conservation assumption implies that the helicity of the virtual photon is preserved by the produced vector meson. As can be seen

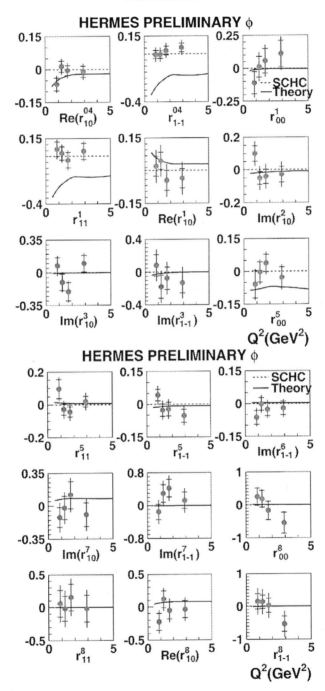

Figure 2. Q^2-dependence of 18 spin density matrix elements for ϕ meson electroproduction on hydrogen; the solid lines represent the calculations of [12], while the dashed lines represent the expected values if s-channel helicity conservation and natural parity exchange are valid.

from Fig. 4 it is worth mentioning that the value of $r_{00}^5 = 0.121 \pm 0.037$ deviates s-channel helicity conservation and is in agreement with the H1 ($r_{00}^5 = 0.090 \pm 0.034$) and ZEUS ($r_{00}^5 = 0.092 \pm 0.031$) measurements [11]. The r_{00}^5 matrix element is proportional to the interference between the helicity conserving longitudinal amplitude T_{00} and the single-flip amplitude T_{01}. The latter amplitude corresponds to the production of longitudinally polarized mesons from transverse photons.

From the measured spin density matrix elements, assuming natural parity exchange in the t-channel, five independent helicity amplitudes have been extracted, which are $T_{00}, T_{11}, T_{01}, T_{10}$ and T_{1-1}. The amplitudes are characterized by their norm $\|T_{ij}\|$ and phase ϕ_{ij}. The parameters $|T_{ij}|, \phi_{10}, \phi_{01}$, and ϕ_{1-1} have large statistical errors and are found to be compatible with zero. In Fig. 5, the results of a fit of the relative size of the remaining parameters are shown under the assumption that $T_{00} = 1$ and $\phi_{00} = 0$. The calculations from [12, 13] are seen in reasonable agreement with the relative sizes of the production amplitudes, but are less successful in describing the phase space. The strong decrease of T_{11}/T_{00} with Q^2 reflects the dominance of the longitudinal over the transverse vector meson production amplitude at large Q^2. The T_{01}/T_{00} ratio of $15.1 \pm 3.5\%$ corresponds to a significant violation of s-channel helicity conservation due to the production of longitudinally polarized vector mesons from transverse photons. These results are in good agreement with data from H1 and ZEUS [14]. Similar helicity violating effects have been reported earlier in [15].

The ratio R of longitudinal over transverse production cross sections has been determined as well using the data for r_{00}^{04}: $R = \sigma_L/\sigma_T = r_{00}^{04}/\epsilon(1-r_{00}^{04})$, where ϵ represents the virtual-photon polarization parameter. The results are compared to existing data for both ρ^0 and ϕ production shown in Fig.4, demonstrating that there is good agreement between low energy and high energy data over a W range of about 100 GeV. Setting r_{00}^5 equal to zero would change the value of R by much less than standard deviation of the data.

4. Summary

From the analysis of vector meson production cross sections at HERMES it has been shown that ρ^0 mesons at intermediate energies are predominantly produced through the quark-exchange mechanism [2]. This conclusion is supported by measured non-zero double spin asymmetry for ρ^0 virtual photoproduction which is in agreement with predictions based on diquark-exchange. These data differ from the results obtained at higher energy experiments.

The spin transfer from virtual photons to vector mesons has been stud-

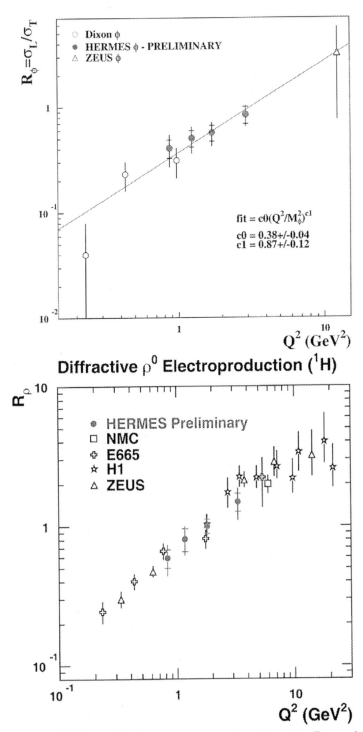

Figure 3. Data on the longitudinal to transverse cross section ratio $R = \sigma_L/\sigma_T$ plotted as a function of Q^2 for ϕ meson production (top panel) and ρ^0 production (bottom panel).

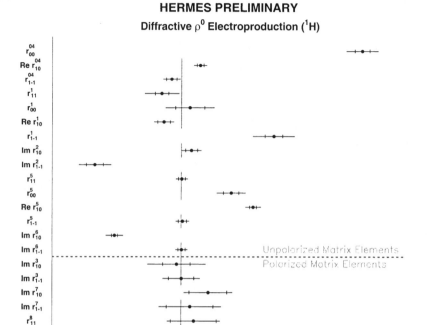

Figure 4. The 23 measured ρ^0 spin density matrix elements. In the top part the 15 elements that can be determined in unpolarized scattering are shown; the bottom are the 8 elements that require longitudinal beam polarization are shown. The broken line represents the values of spin density matrix elements in the case of s-channel helicity conservation.

ied via the measurements of spin density matrix elements. There is good agreement with high energy data on those observed values, on the violation of s-channel helicity conservation in the r_{00}^5 matrix element, on the size of the helicity amplitudes, and the dependence of R on Q^2. Several theoretical calculations [12, 13] describe the spin density matrix elements at high energy, but no analogous calculations are presently available at intermediate energy.

I would like to thank my colleagues of the HERMES collaboration for their advice and support. I am especially grateful to those colleagues whose analysis results have been used in the present paper: K.Lipka, F.Meissner, G.Rakness, and M.Tytgat.

References

1. HERMES Collab. K. Ackerstaff et al, NIM A **417**, 230 (1998).

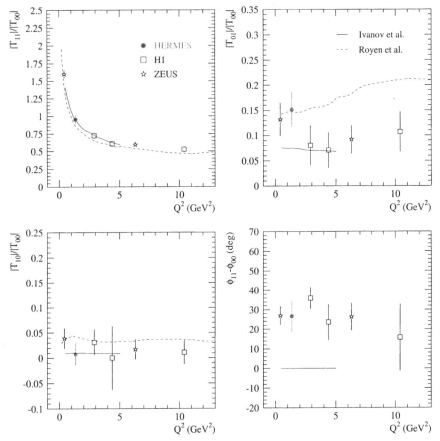

Figure 5.   The helicity amplitude ratios and phase difference $\phi_{11} - \phi_{00}$ as a function of Q^2 are presented. The data are compared to the ZEUS and H1 results [14] and the calculations from [12, 13].

2.   HERMES Collab. K. Ackerstaff et al, Eur.Phys.J. C **17**, 389 (2000).

3.   HERMES Collab. A. Airapetian et al, Phys.Lett. **B 513**, 301 (2001).

4.   E. Lipka, PhD Thesis, Humbold University in Berlin, DESY-THESIS-2002-018.

5.   G.L. Rakness, PhD Thesis, University of Colorado, unpublished (2000).

6.   M. Tytgat, PhD Thesis, University of Gent, DESY-THESIS-2001-018.

7.   H. Fraas, Nucl.Phys. **B113** 532 (1976).

8.   A. Tripet, Nucl.Phys. **B Proc.Suppl.) 79** 529 (1999).

9.   K. Schilling and G. Wolf, Nucl.Phys. **B61** 381 (1973)./

10. H1 Collab. C. Adloff et al., Phys.Lett. **B483** 360 (2000).

11. ZEUS Collab. J. Breiweg et al., Eur.Phys.J. C **12**, 393 (2000).

12. I. Ivanov and N, Nikolaev, JETP Lett., 69 (1999); I. Ivanov priv. comm. (2000).

13. I. Royen, Phys.Lett. **B513** 337 (2001).

14. B. Clerbaux, hep-ph/9908519.

15. W.D. Shambroom et al, Phys.Rev. **D26** 1 (1982); P.Joos et, Nucl.Phys. **B113** 53 (1976).

A FUTURE ELECTRON ION COLLIDER

A. BRUELL

Massachusetts Institure for Technology
77 Massachusetts Avenue, Cambridge, MA 02139, USA

1. Introduction

The Electron-Ion-Collider has been proposed to study in detail the fundamental structure of matter and the theory of the strong interaction, QCD. Despite large experimental and theoretical efforts over the past decades, several of the fundamental questions in this area remain open or incomplete. The proposed machine would collide electrons (positrons) and protons or nuclei at high luminosities and center-of-mass energies between 15 and 100 GeV. Both the electron and the proton beams would be polarised, thus giving acess to the spin struture of the nucleon. Fig. 1 shows the main parameters of the EIC in comparison with existing and planned facilities.

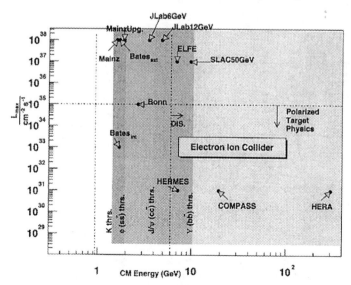

Figure 1. Luminosity vs center-of-mass energy of various existing and proposed facilities

211

E. Steffens and R. Shanidze (eds.), Spin Structure of the Nucleon, 211–219.

The high center-of-mass energy of $\sqrt{s}$ of 15-100 GeV and the high luminosity of $> 10^{33}$ cm^{-2} s^{-1} of the proposed machine will allow to study a large variety of open questions in QCD. The collider geometry will offer a unique possibility to study the final state hadrons and the fragmentation process. In this report only a small selection of these topics can be presented.

2. The Spin Structure of the Nucleon

After three decades of deep-inelastic scattering experiments, the unpolarised structure function of the proton has been measured with enormous precision over four orders of magnitude in x and Q^2. Despite large experimental efforts, the knowledge of the polarised structure functions is still limited (fig. 2). As a consequence the contribution of gluons and sea quarks to the nucleon spin is still unclear. Most important is the limited coverage of the kinematic range which does not yet permit an accurate determination of the polarised gluon distribution from the Q^2 evolution of g_1^p. Also, little is known about the behaviour of g_1^p at low x and moderate Q^2 where a dramatic change towards negative values is expected. The large kinematic range of the proposed EIC (fig. 3) will significantly improve the knowledge of g_1^p especially at low x (see fig. 4).

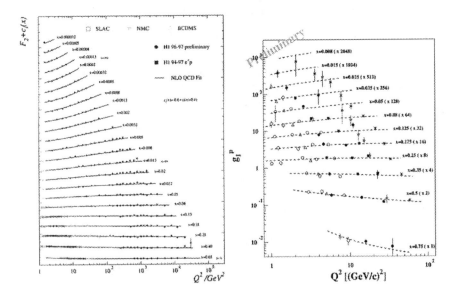

Figure 2. Comparison of the present knowledge of the unpolarised structure function $F_2(x, Q^2)$ and the polarised structure function $g_1(x, Q^2)$ of the proton

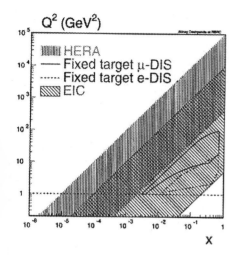

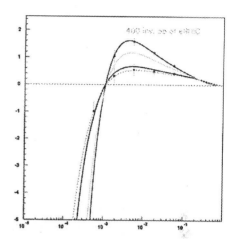

Figure 3. The $x - Q^2$ range of the proposed electron-ion-collider in comparison with previous fixed target experiments and that covered by the unpolarised HERA collider

Figure 4. The expected precision for the polarised structure function $g_1^p(x)$ for 10 GeV electrons and 250 GeV protons at a luminosity of 400 pb^{-1}

Semi-inclusive deep-inelastic scattering has been the tool to access the polarised quark distributions of different flavours. The production of specific hadrons in the final state is correlated with the flavour of the struck quark and allows to determine the contribution of different quarks to the nucleon spin. The most precise data from this method have been obtained by HERMES [1], additional data at higher Q^2 are expected from COMPASS [2] in the near future. A different approach will be used at RHIC-Spin [3] where the parity violation in W^+ and W^- production will allow a precise determination of the polarisations of u and d sea and valence quarks. However, the polarisation of strange quarks will remain poorly known. At the proposed EIC these measurements will be extended into the low x region over a large range in Q^2. Additionally, the EIC will allow a precise measurement of all fragmentation functions, a necessary prerequsit for a determination of the strange quark polarisation via kaon production (see fig. 5).

One of the most interesting questions in spin physics is the question if and how much the gluons contribute to the nucleon spin. Present experiments hardly constrain this quantity and only suggest that this contribution might be positive and sizeable. In the near future measurements at STAR and PHENIX are expected to determine the polarised gluon distribution at high Q^2 but the interpretation of these data might be difficult. At an electron-ion-collider the polarised gluon distribution will be determined by various methods. Fig. 6 shows the expected precision from an analyses of dijet events, a method which has been used successfully to determine the

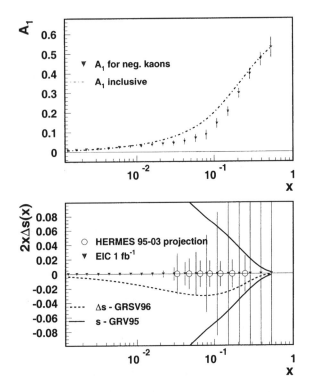

Figure 5. Statistical accuracy for the semi-inclusive kaon asymmetry and the extracted strange quark polarisation for a luminosity of 1 pb^{-1} at the electron-ion-collider

unpolarised gluon distribution at HERA. Additionally, open charm production will give an independent measurement of ΔG at Q^2 values around 10 GeV^2. At these relatively low Q^2 values the maximum of the polarised gluon distribution is expected to be inside the accessible x range of $0.01 < x < 0.3$ resulting into a more precise determination of the integral over $\Delta G(x)$ (fig. 7).

Recently, it has been realised that the structure of the nucleon is described not by two but by three twist-2 structure functions. Beyond the usual unpolarised and longitudinally polarised functions f_1 and g_1, the distribution of transversely polarised quarks is given by a chiral-odd function h_1. Because of its chiral-odd nature this function can not be measured in inclusive scattering but e.g. in semi-inclusive pion production where also the fragmentation function can be chirally odd. A first indication that the h_1 distributions are sizeable has been found at HERMES [4] where a measurement with a transversely polarised target will allow a first determination of the u-quark transversity distribution. However, a precise measurement

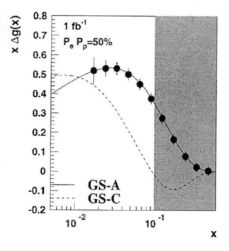
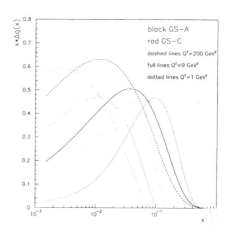

Figure 6. Projected accuracy for the polarised gluon distribution from an analysis of dijet events for a luminosity of 1 pb^{-1} at the electron-ion-collider

Figure 7. The polarised gluon distribution $x\Delta g(x)$ at $Q^2 = 1, 9$ and 200 GeV

of this quantity will require higher luminosities and a larger Q^2 range than available at HERMES. At an electron-ion-collider such a precision measurement would be easily possible.

3. Exclusive Processes

A significant developement over the last years is the identification of a new class of parton distributions, the generalised (or skewed or off-forward) parton distributions (GPD). Probed in exclusive measurements, these GPD's describe hard scattering processes which involve correlations between partons and offer an exciting bridge between elastic and deep-inelastic scattering: in different kinematic limits of the GPD's one recovers the familiar elastic form factors and deep-inelastic structure functions of the nucleon. Further, GPD's have a direct connection to the unknown parton orbital angular momentum which might be an essential contribution to the total spin of the nucleon.

One of the reactions that can be most clearly interpreted in terms of GPD's is deeply-virtual Compton scattering (DVCS), i.e. the production of one high-energy photon with the target nucleon staying intact. First measurements of the beam-spin asymmetry in hard exclusive electroproduction of photons have been reported by the HERMES and the CLAS collaborations [5] (fig. 8). Using the presently unique capability of the HERA accelerator to deliver both electrons and positrons, HERMES also reported a first measurement of the beam-charge asymmetry asociated with the DVCS process [6]. However, the kinematic range at CLAS and at HERMES (especially the limited Q^2 range) will allow only a very first glimps at this interesting process while a collider with high luminosity and high center-of-mass energy would cover a large range in x and Q^2 (fig. 9).

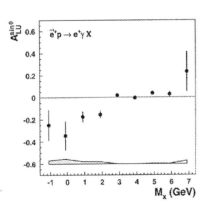

Figure 8.   First measurement of the beam-spin asymmetry A_{LU} for hard exclusive electroproduction of photons as a function of the missing mass M_X

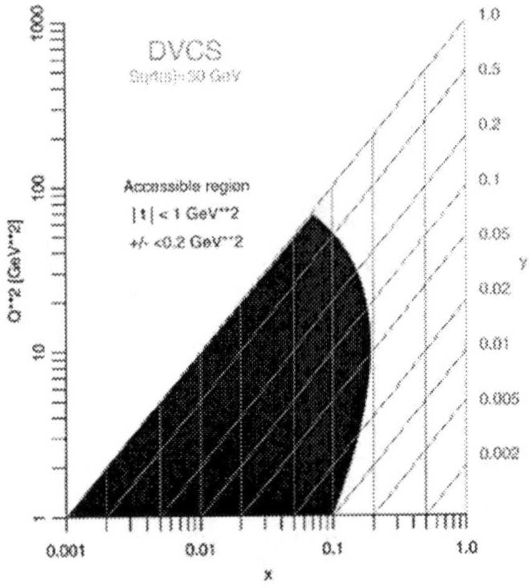

Figure 9.   The $x - Q^2$ range for the DVCS process for a center-of-mass energy of $\sqrt{s} \sim 30$ GeV

4.   Nuclear Effects in Deep-Inelastic Scattering

Since the discovery of the EMC effect [7], parton distributions are known to be modified in the nuclear medium compared to the free nucleon case. Subsequent experiments have measured the $x-$, Q^2- and A-dependence of the nuclear effects in deep-inelastic scattering with high precision [8]. However, the interpretation of these data is not . However, the interpretation of these data is not unique: as charge lepton scattering probes the gluon distribution only indirectly through the Q^2 evolution of the measured structure functions, the observed shadowing in the nuclear structure function can be modeled by different combinations of quark and gluon distributions in the nucleus. In particular, an analysis of the most precise data available, the NMC measurement of F_2^{Sn}/F_2^{C} [9], resulted into a possible enhancement of the gluon distribution in Sn compared to that in C at intermediate x values of up to 40% [10].

At a collider with a high center-of-mass energy and the option to accelerate light and heavy nuclei, measurements of structure function ratios

on a nucleus A with respect to deuterium could improve our knowledge
of these medium modifications significantly. Of special importance here is
the extension of the measurement to low values of x over a large Q^2 range.
Additionally, an electron-ion-collider would allow a detailed study of the
nuclear dependence of diffractive processes and jet production. Both processes are sensitive to the gluon distribution in the nucleus. Understanding
the origin of shadowing is of the highest importance for the interpretation
of any high-energy process involving nuclei. Shadowing influences the final
state particle multiplicities and energy densities in relativistic heavy ion
collisions, results into opacity effects that can mask color transparency in
vector meson production and limits the present ability to calculate partonic
energy loss in the nuclear medium.

5. High Density Partonic Matter

A high energy electron-nucleus collider presents a remarkable opportunity
to study a very fundamental aspect of QCD. The nucleus acts as an am-
plifier for the novel physics expected at high parton densities - aspects of
QCD which otherwise could only be explored in an electron-proton collider
with energies well above the HERA collider at DESY.

As we go to higher energies, the density of small x partons grows as $x^{-\delta}$
with $\delta \sim 0.3$ for Q^2 values of a few GeV^2. As a high energy probe simultaneously resolves partons from different nucleons along its trajectory, this
density also grows as $A^{1/3}$. The density per unit rapidity associated with
these partons is $Q_S^2 = A^{1/3}x^{-\delta}$. If Q_S is big enough, the strong interaction is expected to become a weak interaction ($\alpha_S << 1$) and thus a scale
invariant theory. In this limit, many features of multi-particle production
can be computed from first principles. Fig. 10 shows a schematic view of
the $\ln(1/x)$ and $\ln(Q^2/\Lambda^2)$ plane. In the region of either high Q^2 or large x
where the desnity of partons is moderate, the evolution of parton densities
with respect to Q^2 or $\ln(1/x)$ is usually described by the DGLAP or BFKL
evolution equations. Evolving into the high parton density regime (yellow
area), these linear evolution equations should get corrections resulting into
a moderation or even saturation of the growth of the parton densities. Experimentally, there is evidence from pA experiments at Fermilab that the
typical momenta of partons increases as $p_T^2 \sim Q_S^2$. There are also hints from
HERA that one may be beginning to see high parton density effects at Q^2
values around 1-10 GeV and the lowest available values of x. Phenomenological models that contain a saturation scale $Q_S^2 \sim 1$ GeV^2 at $x \sim 10^{-4}$
have been sucessful in explaining the HERA inclusive, diffractive and vector meson production data. The same models predict a saturation scale of
about 5-10 GeV for the electron-ion-collider at similar x values.

A number of inclusive and semi-inclusive observables are expected to be sensitive to the effects of high parton densities. As already mentioned, at the luminosities envisaged at the EIC, the structure function $F_2(x, Q^2)$ and its logarithmic derivative can be measured with high accuracy. Corrections due to high parton densities will result into deviations from pQCD fits incorporating the standard DGLAP evolution. The same feature is expected in the longitudinal structure function $F_L = F_2 - 2xF_1$. Fig. 11 shows the prediction for the ratio of the longitudinal to transverse structure functions which should show a maximum at a particular Q^2 related to the saturation scale Q_S^2. Other quantities sensitive to high parton density effects include the fraction of diffractive events and the cross section for coherent vector meson production.

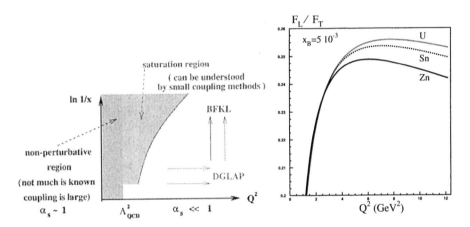

Figure 10. Schematic view of the $ln(1/x) - Q^2$ plane showing the regimes of applicability of different evolution equations

Figure 11. The ratio F_L/F_T as a function of Q^2 at fixed x for different nuclei

6. Possible Accelerator Oprtions

Several collider configurations are considered to achieve the technically challenging goal of a luminosity of about 10^{33} cm^{-2} s^{-1}. One scenario uses the proton (ion) ring at RHIC together with a new small electron ring. Others would use the same hadron ring but a linac to accelerate the electrons. Other options as using the JLAB electron machine are also considered. In all cases significant R&D effort is required.

References

1.   HERMES, K. Ackerstaff et al., Phys. Lett. B464 (1999) 123;
     M. Beckmann for the HERMES collaboration, hep-ex/0210049.

2. COMPASS, http://wwwcompass.cern.ch/compass/proposal.
3. Polarised Proton Collider at RHIC, http://www.agsrhichome.bn.gov/RHIC/spin
4. HERMES, A. Airapetian et al., Phys. Rev. Lett. 84 (2000) 4047.
5. HERMES, A. Airapetian et al., Phys. Rev. Lett. 87 (2001) 182001; CLAS, S. Stepa-nyan et al., Phys. Rev. Lett. 87 (2001) 182002.
6. F. Ellinghaus (for the HERMES collaboration), hep-ex/0207029.
7. EMC, J.J. Aubert et al., Phys. Lett. B123 (1983) 275.
8. SLAC, J. Aubert et al., Phys. Lett. B123 (1983) 275; NMC, P. Amaudruz et al. (NMC), Nucl. Phys. B441 (1995) 3; NMC, M. Arneodo et al. (NMC), Nucl. Phys. B441 (1995) 12.
9. NMC, M. Arneodo et al., Nucl. Phys. B481 (1996) 23.
10. T. Gousset, H.J. Pirner, Phys. Lett. B375 (1996) 349.

POLARIZATION OBSERVABLES IN THE N→Δ(1232) TRANSITION

SALVATORE FRULLANI
INFN,Gruppo Sanità and Istituto Superiore di Sanità,Physics Laboratory - Rome, Italy

JAMES J. KELLY
University of Maryland, Department of Physics - College Park, MD 20742, U.S.A.

ADAM J. SARTY
Saint Mary's University,Department of Astronomy & Physics - Halifax, NS, Canada

ROBERT W. LOURIE
Renaissance Technologies Corporation, East Setauket, NY 11733-2841, U.S.A.

MARK K. JONES AND DAVE G. MEEKINS
Thomas Jefferson National Accelerator Facility, Newport News, VA 23606, U.S.A.

WILLIAM BERTOZZI, ZHENGWEI CHAI AND SHALEV GILAD
M.I.T., Department of Physics - Cambridge, MA 02139, U.S.A.

RIKKI E. ROCHÉ
Florida State University, Physics Department - Tallahassee, FL 32306, U.S.A.

AND

THE JEFFERSON LAB HALL A COLLABORATION

1. Introduction

In the simple SU(6) spherically symmetric quark model, the proton and the Δ(1232) are states of three quarks in an overall S-wave with spins coupled to 1/2 and 3/2 respectively. In this context, the transition has a pure magnetic dipole M_{1+} nature resulting in the spin-flip of a single quark. However the measurements carried out in the late sixties and early seventies at DESY [1, 2, 3], NINA (Daresbury) [4] and CEA (Harvard) [5] have demonstrated

E. Steffens and R. Shanidze (eds.), Spin Structure of the Nucleon, 221–234.

already the inconsistency of the data with a pure magnetic dipole transition, indicating that, although small, other components were also needed.

Due to selection rules, the only other resonant multipole contributions to the excitation of the $\Delta(1232)$ are the Coulomb quadrupole S_{1+} and the electric quadrupole E_{1+}. These quadrupolar transitions imply non-spherical forces - whose dynamical origin is ascribed to different mechanisms by the various models (see as examples [6, 7, 8]) - which could result in a L=2 component in the nucleon or/and delta wave functions. This problem is obviously of interest for the spin physics studied in polarized deep inelastic scattering in connection with the question of the contribution of quarks and gluons orbital angular momenta to the nucleon spin.

In addition to the resonant transitions other coherent nonresonant processes and tails of transitions to higher energy resonances contribute as background to the yield under the Δ peak . These processes must be determined or at least constrained with measurements and model computations in order to extract in a reliable way the strength of the S_{1+} and E_{1+} contributions to the N$\rightarrow\Delta(1232)$ transition.

Due to the smallness of other components with respect to the magnetic dipole transition the most useful way to quantify their importance is through the simultaneous measurement of observables that depend linearly on interference terms of one of the small components with the dominant (M_{1+}).

The previously mentioned old experimental data, resulting from measurements carried out with unpolarized, low duty cycle and low intensity beams, allowed the determination of a small number of observables with large uncertainties and with the consequent unsatisfactory knowledge about the quadrupolar components. The new facilities like MAMI, ELSA, LEGS, Bates-MIT and Jlab have renewed, in recent years, the interest on the subject motivating measurements of high quality.

In figure 1 most of the available data on the extracted Coulomb and electric quadrupole terms are shown. The situation is sensibly improved with new results in very recent years both in photo and electro-production. However, due to the complexity of data reduction, results obtained from measurements using different techniques have to be compared in order to fully size systematic and model dependent uncertainties.

The availability of a Focal Plane Polarimeter (FPP) to measure recoil proton polarization as standard experimental tool in Hall A at Thomas Jefferson National Accelerator Facility motivated the E91-011 proposal to study the N$\rightarrow\Delta(1232)$ transition through the measurement of polarization observables. Merits of this technique can be found in ref. [9, 10] and can be summarized as giving access to a simultaneous measurement of many

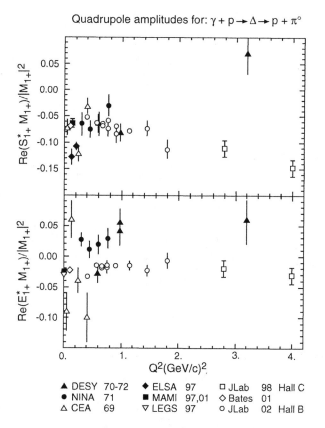

Figure 1. Q^2 dependence of the Coulomb (S_{1+}) and electric (E_{1+}) quadrupole/magnetic dipole ratios. Only statistical errors are shown. Data points are from DESY [1, 2, 3], NINA [4], CEA [5], ELSA [11], MAMI [12, 16], LEGS [13], Jlab [14, 15], Bates [18].

Response Functions (see next section) and to the determination of both real and imaginary parts of interference amplitudes.

This last feature is of particular importance when resonant and non-resonant parts contribute to the yield at the peak of a resonance. Scattering amplitudes can be decomposed in multipole amplitudes and all observables can be written as a sum of bilinear products of multipoles amplitudes. At the resonance peak, resonant amplitudes are purely imaginary and background amplitudes are essentially real, then the interference term with the M_{1+} amplitude has only a real part for a small resonant amplitude and only an imaginary part for a non resonant amplitude. These constraints allow, in principle, resonant and non-resonant contributions to be characterized.

The technique has already been used at lower momentum transfer in two experiments [16, 17] that have carried out the measurement only in parallel kinematics, when the proton is detected along the direction of momentum transfer. The E91-011 experiment has covered a large fraction of the kinematical phase space and the technical possibility of the extraction of the Response Functions (RFs) through recoil polarization measurements can be fully tested.

2. Extraction of Response Functions from Polarization Observables

Studying exclusive or semi-exclusive (semi-inclusive) reactions induced by polarized charged leptons with the detection, in coincidence with the scattered lepton, of one other final state particle, when the polarization of the target or of the detected outgoing particle is determined, allows the extraction of a wealth of information on the characteristic of the transition between the initial and the final states of the system under study. This will be discussed specifically for the reaction studied and afterwards some considerations will be added for a more general case.

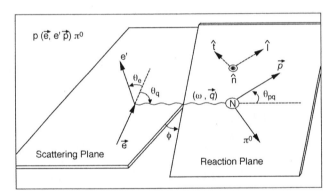

Figure 2.   Kinematical variables relevant for the reaction under study. Capability to detect out-of-plane reactions is determined by the ϕ (azimuthal angle) acceptance of the detection apparatus at different θ_{pq} values.

In figure 2 some kinematical notations relevant for the scattering process considered are shown.

The cross section for the $p(\vec{e}, e'\vec{p})\pi^0$ process follows the general expression for a reaction induced by a polarized electron off an unpolarized target and when the polarization of the outgoing proton is measured that is given by [19, 20]:

$$\frac{d^5\sigma}{de'\,d\Omega_e\,d\Omega_p} \approx \sigma_0[1 + \mathbf{P} \cdot \mathbf{S} + h(A_e + \mathbf{P'} \cdot \mathbf{S})] \tag{1}$$

where σ_0 is the five-fold unpolarized differential cross section, $\mathbf{S}$, $\mathbf{P}$ and $\mathbf{P'}$ are the spin, the induced and transferred polarizations of the outgoing proton, respectively, h is the beam helicity and A_e is the beam analysing power. Equation (1) refers to an exclusive process but a similar relation, with the cross sections now six-fold differential (adding the proton momentum differential), also holds for semi-exclusive (semi-inclusive) processes.

In figure 3 the relations between the measurable quantities of equation (1) and the Response Functions are shown.

18 Response Functions

$$\sigma_0 = v_L \boxed{R_L} + v_T \boxed{R_T} + v_{LT} \boxed{R_{LT}} \cos \phi + v_{TT} \boxed{R_{TT}} \cos 2\phi$$

$$P_n \sigma_0 = v_L \boxed{R_L^n} + v_T \boxed{R_T^n} + v_{LT} \boxed{R_{LT}^n} \cos \phi + v_{TT} \boxed{R_{TT}^n} \cos 2\phi$$

$$P_{l,t} \sigma_0 = v_{LT} \boxed{R_{LT}^{l,t}} \sin \phi + v_{TT} \boxed{R_{TT}^{l,t}} \sin 2\phi$$

$$A_e \sigma_0 = v'_{LT} \boxed{R'_{LT}} \sin \phi$$

$$P'_n \sigma_0 = v'_{LT} \boxed{R'^n_{LT}} \sin \phi$$

$$P'_{l,t} \sigma_0 = v'_{LT} \boxed{R'^{l,t}_{LT}} \cos \phi + v'_{TT} \boxed{R'^{l,t}_{TT}}$$

How to separate:

handle	Real type	Imaginary type
Left/Right	$\boxed{R_{LT}}$, $\boxed{R_{LT}^{l,t}}$, $\boxed{R_{TT}^{l,t}}$	$\boxed{R_{LT}^n}$
Out of plane	$\boxed{R_{TT}}$, $\boxed{R_{LT}^{'n}}$, $\boxed{R_{TT}^n}$	$\boxed{R'_{LT}}$, $\boxed{R_{LT}^{'l,t}}$, $\boxed{R_{TT}^{'l,t}}$
Rosenbluth	$\boxed{R_L}$, $\boxed{R_T}$	$\boxed{R_L^n}$, $\boxed{R_T^n}$

Figure 3. Relations between measured quantities and Response Functions. Methodology to extract individual Response Functions.

The ν are factors dependent on the electron kinematical variables, ϕ is the angle between scattering and reaction planes, the subscripts L,T,LT and TT refer to the longitudinal, transverse components of the hadronic tensor and interference between longitudinal and in plane transverse components and interference between in plane and orthogonal transverse components of the hadronic tensor respectively, all referring to the scattering plane and direction of the three-momentum transfer, the indexes l,n,t refer to the proton polarization components according to the reference system of figure 2, while unprimed and primed refer to beam helicity independent and dependent terms, respectively.

RFs are dependent on four scalar arguments that are usually chosen as the quadrimomentum transfer Q^2, the center-of-mass energy W, recoil proton momentum p and emission angle of the recoil proton with respect to the three momentum transfer θ_{pq}.

In figure 3 the methodology to single out the different RFs is also indicated. For fixed values of the arguments of the RFs, terms containing $\cos \phi$ allow extraction with a substraction from measurements carried out in plane at left/right of three momentum transfer ($\phi = 0^o$ and 180°), terms containing $\sin \phi$, $\sin 2\phi$ or $\cos 2\phi$ need instead combination of out-of-plane ($\phi \neq 0^o$ or 180°) measurements. All the mentioned measurements are carried out at the same kinematical values of incident and scattering electron variables, while for longitudinal and transverse RFs a Rosenbluth separation is required with different values of ν_L and ν_T obtained changing electron kinematics. Unprimed and primed RFs are separated combining measurements with opposite beam helicity.

RFs are real or imaginary parts of bilinear products of amplitudes and from the inset table in figure 3 is evident how polarization measurements are essential to determine both parts.

For a process, involving n particles with intrinsic spin j_i, in which the parity is conserved, the number of independent scattering amplitudes is given by $N = \prod_{i=1}^n (2j_i + 1)/2$, resulting in 2N-1 real parameters with one overall phase undeterminable. In the limit of the one photon (j=1) exchange approximation, our reaction is then described by 6 linearly independent complex scattering amplitudes and 11 real parameters which can be determined with opportunely chosen measurements [19, 21] which give the maximum obtainable knowledge of the reaction.

Naturally the determination of as many as possible RFs gives constraints on the derived scattering amplitudes.

A similar strategy to extract RFs can be applied if, instead of measuring the polarization of the outgoing proton, the polarization of the target is defined and the asymmetries of the reaction are measured when the target spin is flipped in different directions. In this case, analogous to equation (1), the cross section is given by [22, 23]:

$$\frac{d^5\sigma}{de' d\Omega_e d\Omega_p} \approx \sigma_0 [1 + \mathbf{A} \cdot \mathbf{S} + h(A_e + \mathbf{A}' \cdot \mathbf{S})] \qquad (2)$$

where $\mathbf{S}$, $\mathbf{A}$ and $\mathbf{A}'$ are the vector polarization of the target, the target analysing power independent and dependent (spin correlation coefficient in the more usual jargon of polarization experiments) respectively, from the beam helicity. The components of these last two vectors are determined through asymmetries measurements of the target polarized in the direction of the axis of the reference system with unpolarized and polarized beam.

Relations analogous to those of figure 3 hold for helicity independent and dependent target analysing power components and the 18 RFs that are now referred as target polarization RFs. The two sets of RFs - target polarization and recoil polarization - are clearly related to each other.

It has been shown [21] that to have the complete determination of the 11 parameters mentioned above only one set of measurements is not enough without at least one experiment from the other polarization set, while experiments with both target and recoil polarization are not necessary for the complete determination of the process.

The number of RFs that is required in the description of a reaction induced by a charged lepton beam depends on the spin of the polarized target or of the detected particle whose polarization is measured, for instance for a spin 1 particle the total number of RFs becomes 33 [9] instead of the 18 for a spin 1/2 particle.

As mentioned, the number of linearly independent scattering amplitudes depends on the spins of all the particles in the initial and final states of the reaction. For deuteron electrodisintegration or for Virtual Compton Scattering off a proton this number is 18, which implies the measurements of 35 selected parameters [24].

The structure of the cross section according to equations (1,2) and its dependence on 18 RFs derives from very general invariance properties of the hadronic tensor [25, 26] such as hermiticity, current and parity conservation. Then, it holds for every reaction induced by a longitudinal polarized charged lepton beam diffusing off a polarized spin 1/2 target with the detection of one hadron in the final state in coincidence with the diffused lepton or when the diffusion is on an unpolarized target with arbitrary spin but the detected hadron is a 1/2 spin particle and its polarization is measured. Independently from the energy scale of the reaction, the general formulation is valid in the region of single levels nuclear excitation, giant dipole resonance, quasifree nucleon scattering, resonances electroproduction and also for exclusive or semi-inclusive deep inelastic (SIDIS), an energy regime more in the focus of this workshop.

In SIDIS regime models describe the reaction at the parton level. The simplest diagram representing the squared amplitude at leading order is shown in figure 4a. The hadronic tensor contains two soft parts represented by the blobs in the figure, these parts contain the relevant information on the hadron structure at the parton level and on the quark-hadron dynamic. Φ contains information on how the momentum and spin of the quarks is distributed in the target hadron (distribution functions), Δ is the correlator giving the production probabilities of the final hadron from the quark that has been scattered in the hard process with the virtual photon (fragmentation functions).

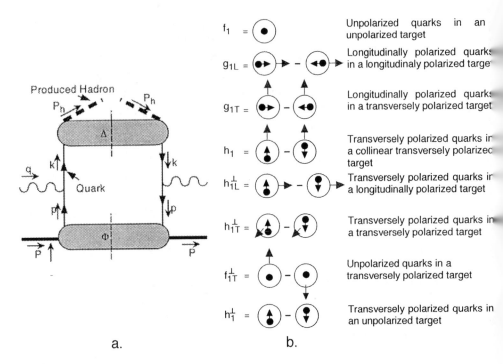

f_1 = Unpolarized quarks in an unpolarized target

g_{1L} = Longitudinally polarized quarks in a longitudinaly polarized target

g_{1T} = Longitudinally polarized quarks in a transversely polarized target

h_1 = Transversely polarized quarks in a collinear transversely polarized target

$h_{1L}^{\perp}$ = Transversely polarized quarks in a longitudinally polarized target

$h_{1T}^{\perp}$ = Transversely polarized quarks in a transversely polarized target

$f_{1T}^{\perp}$ = Unpolarized quarks in a transversely polarized target

$h_1^{\perp}$ = Transversely polarized quarks in an unpolarized target

a.   b.

Figure 4. Semi-inclusive scattering process (a.) and leading order distribution functions (b.). Adapted from [27].

In figure 4b the leading order (twist two) distribution functions are shown. They appear as combinations of measurements with unpolarized and polarized targets with different polarization directions and they are interpreted as parton momentum or spin densities in unpolarized or polarized targets [27]. Other distribution functions appear at higher subleading orders but are not interpretable as densities or differences of densities. A description of the process up to twist three is available [28]. A complete strategy to single out the functions of interest has not been found yet.

SIDIS induced by polarized leptons off polarized protons can be described by a cross section of the type of equation (2), while the study of semi-inclusive Λ production from an unpolarized target with determination of its polarization through the self analysing power of its weak decay is described by equation (1). As said, in both cases a well established strategy exists to extract single RFs. Then, particularly important is a more transparent deduction, in the framework of SIDIS models, of the Response Functions in terms of distribution and fragmentation functions in order to evaluate if the methodology of extraction adopted can be of help also to single out the functions of interest in SIDIS regime, sharing a common practise and language with other related fields of lepton scattering.

3. E91-011 Measurements

The experiment has been carried out in Hall A at JLab [29], utilizing its two high resolution spectrometers. The strained GaAs source provided an electron beam with a longitudinal polarization that remained within 67 to 79% for all the time of data taking (35 days) at an average current of 45 μA, with the beam helicity flipping pseudorandomly at 30 Hz. The polarization was monitored continously by a Compton [30] and occasionally by a Møller polarimeter. The two measurements were in agreement within a few percent. A liquid H_2 target 15 cm long was used.

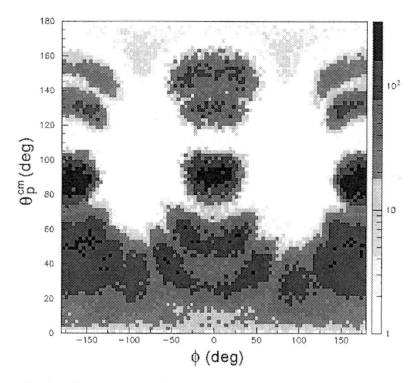

Figure 5. Angular coverage for the measurements. Contours of normalized yield are shown for (θ_p^{cm}, ϕ), combining acceptances of the various kinematical settings.

The electron kinematics has been kept fixed for all the kinematical sets of data production : incident and scattered electron energy = 4.535 and 3.066 GeV respectively, and electron scattering angle = 14.09^0, defining a $Q^2 = 1.0 \ (GeV/c)^2$. Proton momenta and angles were chosen to have an invariant mass of the hadronic final system of W = 1.232 GeV and to cover in an adequate way the center-of-mass (COM) angular distribution.

In figure 5 the angular phase space coverage is shown. For all θ_p^{cm} settings, in plane measurements, centered at $\phi = 0^o$ and 180°, were carried out to allow the extraction of 6 RFs with the "left/right" handle. The Lorentz boost limits the window of laboratory angle of the emitted particles and this results in a considerable ϕ acceptance for the settings with small values of θ_p^{cm}, allowing the extraction of other 8 RFs with the "out-of-plane" handle. For the other 4 RFs, a separated combination of longitudinal and transverse RFs is extracted for unpolarized and for n component induced polarization RFs. In addition, the measurement in parallel kinematics ($\theta_p^{cm} = 0^o$) will allow the determination of R_L/R_T at that angle [31].

The Hall A High Resolution Spectrometer (HRS) pair was used for the detection of the electrons and protons. The electron arm included a two-plane scintillator array, providing triggering and time-of-flight information for coincident particle identification, as well as an atmospheric CO_2 threshold Čerenkov detector and a segmented lead glass calorimeter for π^-/e discrimination. In order to track trajectories, two vertical drift chambers each with a pair of wire planes were used. The proton arm included the same system of scintillators and tracking detector and in addition contained a focal plane polarimeter (FPP) [32] consisting of an adjustable thickness graphite analyzer sandwiched by straw tube drift chambers that allowed the measurement of polar and azimuthal angles after scattering in the analyzer.

The determination of the two components of the proton polarization at the FPP, transverse to the momentum, is accomplished through Fourier analysis of the angular distribution, using the known analyzing power of carbon for the energies of interest. Due to the precession of the proton spin in the different magnetic elements of the HRS, the polarization components at the target are different from the values measured at the FPP. They are related through a 3x3 spin transport matrix that depends on the proton trajectory and then is unique for each event. A maximum likelihood method [33] was used to obtain the polarization at the target.

To check the polarimetry analysis, several elastic $p(\vec{e}, e'\vec{p})$ measurements were carried out, with proton momenta spanning the range used in the settings of the experiment. In this way it has been possible to check/calibrate the instrumental asymmetries in the FPP, as well as cross check extracted polarizations with measurements already available.

All the components of the induced and transferred polarization at the target have been determined.

In figure 6, as an example, the azimuthal angular distributions of the n component of the induced polarization at the target for several intervals

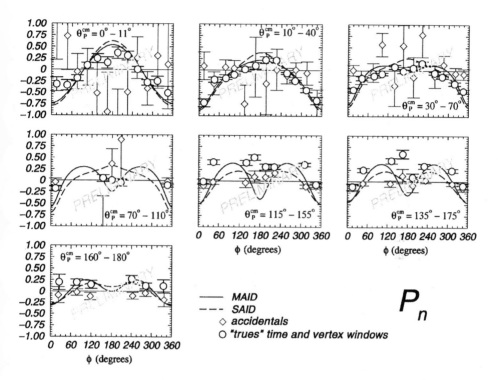

Figure 6. Azimuthal angular distributions of the induced polarization n component at target. Results are compared with SAID [34] and MAID [35] parametrizations.

of the COM angle are shown. Experimental points refer to the measured polarization for accidentals and the "trues", defined in the appropriate co-incidence time and reconstructed target position windows, while curves are the results of computations using scattering amplitudes with parametrizations derived from a phase shift analysis of the world's data (SAID) [34] or from the Mainz unitary isobar model (MAID) [35].

To measure cross sections one must rely on a good knowledge of the phase space acceptance of the experimental apparatus. For all the experiment, selection of events was made applying cuts on the correlations between the reconstructed origin in the target and coincidence time, between the ADC level of the pulse from the scintillators of the hadron arm and the β of the particle and between missing mass and missing momentum. The results of a simulation, using the Hall A Monte Carlo simulator MCEEP with MAID as an event generator, of the spectrometers acceptances for all variables showed remarkable agreement with the shape of the measured distributions.

Azimuthal angular distributions of the unpolarized reduced cross section (σ_0 divided by the flux of the vitual photons) for some values of the pion

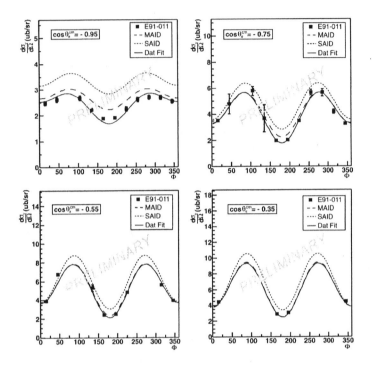

Figure 7. Azimuthal angular distributions of the unpolarized reduced cross section.

COM angle are shown in figure 7. The precision of the data points seems to be able to discriminate between models.

From the measured polarization components and unpolarized cross section it is possible to extract the Response Functions. This is again made using a maximum-likelihood method of the type used before but with the RFs having replaced the polarization components, according to the relations of figure 3, in the function to be minimized.

In figure 8 the extracted helicity-independent polarized RFs are shown. Data are compared with the results of SAID and MAID parametrizations and indicate that the data accuracy is good enough to discriminate between the two models that differ significantly due to the different weight of non resonant components of the amplitude. In the figure it is also shown that MCEEP simulation, that takes properly into account the detector acceptances, does not differ from the results obtained with the same parametrization (MAID) using kinematical values at the center of the acceptances, indicating that the eventual difference between data and a theoretical curve is not due to acceptance problems. In figure 8 , to extract RFs from the polarization data, the computed values (MCEEP with MAID) of the unpolarized cross section were used; work is in progress to use instead the

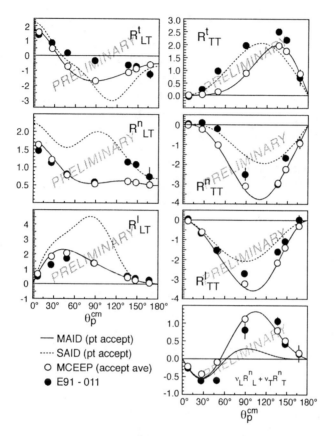

Figure 8. Helicity-independent polarization Response Functions. Results are compared with SAID [34] and MAID [35] parametrizations.

experimental measured values.

Work is also in progress to extract the unpolarized RFs. With all the RFs extracted, the multipole decomposition will be worked out with the aim to precisely quantify all the contributions to the N→Δ(1232) transition.

4. Conclusions and outlook

For the first time the simultaneous separation of many RFs with recoil polarization technique has been applied for measurements covering a large fraction of the phase space, for a single Q^2, W setting.

The technique has been proven valid and 14 RFs have been individually extracted within the same set of measurements with the additional information coming from two combinations of the other 4.

Multipole decomposition, the most model independent analysis that can be made, is still to be done. Its result should give the physical information about the size of the small resonant and non resonant contributions. This is the physics aim of the experiment and will define the sensitivity of the technique. From this result one will infer the suitability to perform measurements at other Q^2, W settings or/and for other reactions.

The technique is applicable to leptoproduction experiments in higher energy regimes provided that models, supposed to describe hadronic systems in that regimes, offer the opportunity to test their predictions in terms of RFs.

This work was supported by DOE contract DE-AC05-84ER40150 under which the Southeastern Universities Research Association (SURA) operates the Thomas Jefferson National Acceleration Facility and in part by contract DE-FC02-94ER40818.

References

1. Albrecht, W. *et al.* (1970) *Nucl. Phys.* **B25**, 1
2. Albrecht, W. *et al.* (1971) *Nucl. Phys.***B27**, 615
3. Alder, J. C. *et al.* (1972) *Nucl. Phys.***B46**, 573
4. Siddle, R. *et al.* (1971) *Nucl. Phys.* **B35**, 93
5. Mistretta, C. *et al.* (1969) *Phys. Rev.* **184**, 1487
6. Isgur, N., Karl, G. and Koniuk, R. (1982) *Phys. Rev.* **D25**, 2394
7. Capstick, S. (1992) *Phys. Rev.* **D46**, 2864
8. Silva, A. *et al.* (2000) *Nucl. Phys.* **A675**, 637
9. Raskin, A. D. and Donnelly, T. W. (1989) *Ann. Phys.* **191**, 78
10. Lourie, R. W. (1992) *Phys. Rev.* **C45**, 540
11. Kalleicher,F. et al. (1997) *Z. Phys.* **A359**, 201
12. Beck, R. *et al.* (1997) *Phys. Rev. Lett.* **78**, 606
13. Blanpied, G. *et al.* (1997) *Phys. Rev. Lett.* **79**, 4337
14. Frolov, V. V. *et al.* (1999) *Phys. Rev. Lett.* **82**, 45
15. Joo, K. *et al.* (2002) *Phys. Rev. Lett.* **88**, 122001
16. Pospischil, Th. *et al.* (2001) *Phys. Rev. Lett.* , **86**, 2959
17. Warren, G. A. *et al.* (1998) *Phys. Rev.* **C58**, 3722
18. Mertz, C. *et al.* (2001) *Phys. Rev. Lett.* **86**, 2963
19. Giusti, C. and Pacati, F. D. (1989) *Nuc. Phys.* **A504**, 685
20. Kelly, J. J. (1996) *Adv. Nuc. Phys.* **23**, 75
21. Dmitrasinovic, V.*et al.* (1987) Research Program at CEBAF (III), 547
22. Laget, J. M. (1991) *Phys. Lett.* **B273**, 367
23. Boffi, S., Giusti, C. and Pacati, F. D. (1993) *Phys. Rep.* **226**, 1
24. Dmitrasinovic, V. and Gross, F. (1989) *Phys. Rev.* **C40**, 2479
25. Gourdin, M. (1972) *Nucl. Phys.* **B49**, 501
26. Picklesimer, A. and Van Orden, J. W. (1987) *Phys. Rev.* **C35**, 266
27. Mulders, P. J. (2002) *Czech. Journ. Phys.* **52 Suppl. A**, A1
28. Mulders, P. J. and Tangerman, R. D. (1996) *Nucl. Phys.* **B461**, 197
29. Anderson, B. D. *et al.* submitted to *Nucl. Instrum. Meth. in Phys. Res.*
30. Baylac, M. *et al.* (2002) *Phys. Lett.* **B539**, 8
31. Kelly, J. J. (1999) *Phys. Rev.* **C60**, 054611
32. Bimbot, L. *et al.* submitted to *Nucl. Instrum. Meth. in Phys. Res.*
33. Wijesooriya, K. *et al.* (2002) *Phys. Rev.* **C66**, 034614
34. Arndt, R. A., Strokovsky, I. I. and Workman, R. L. (1996) *Phys. Rev.* **C53**, 430
35. Drechsel, D. *et al.* (1999) *Nucl. Phys.* **A645**, 145
36. Ulmer, P. www.physics.odu.edu/ulmer/mceep/mceep.html

COHERENT ρ^0 MESON PRODUCTION AT HERMES

A. V. AIRAPETIAN
Yerevan Physics Institute, Alikhanian Brs. st. 2,
Yerevan, 375036, Armenia
E-mail: Avetik.Airapetian@desy.de
ON BEHALF OF THE HERMES COLLABORATION

Abstract. Recent HERMES measurements are presented on the coherent and incoherent part of the cross section for the exclusive diffractive production of ρ^0 mesons. The results are based on data taken with a variety of targets: ^{1}H, ^{2}H,^{3}He, ^{14}N, ^{20}Ne and ^{84}Kr. The method to extract the coherent-to-incoherent cross section ratios as a function of coherence length l_{coh} and momentum transfer Q^2 is presented. The measurement of these ratios can contribute to the study of how quark-antiquark pairs interact with the nuclear medium. Using the ^1H and ^{14}N targets, the nuclear transparency of ρ^0 production has been measured as a function of coherence length. The data have been analysed to search for a possible onset of color transparency and are compared with recent theoretical calculations.

E. Steffens and R. Shanidze (eds.), Spin Structure of the Nucleon, 235–242.

1. Introduction

Exclusive electroproduction of ρ^0 mesons from nuclei is considered to be an excellent tool to investigate the properties of elementary particles interacting with the nuclear medium, such as the phenomena of a "shrinking photon"[1] and Color Transparency (CT)[2]. The latter phenomenon is a prediction of perturbative QCD, and suggests that particles produced with high virtuality in exclusive reactions should exhibit a reduced interaction with other hadrons due to their reduced transverse size. In particular, the "size" of the hadronic components of the virtual photon at high negative four-momentum transfer squared, Q^2, is conjectured to be smaller than the size of a normal hadron. This would account for the pointlike behavior and the diminished absorption of virtual photons in nuclear interactions, as compared to real photons. In QCD, the reaction amplitudes for exclusive interactions at large momentum transfer are expected to be dominated by components of the photon wavefunction with small transverse size, which give rise to diminished final state interactions in the nuclear medium. Theoretical models typically describe the exclusive production of light vector mesons as occuring via the fluctuation of the virtual photon into a quark-antiquark pair (or off-shell vector meson), which is scattered onto the mass shell by a diffractive interaction with the target. The corresponding tree level diagram is shown in Fig. 1.

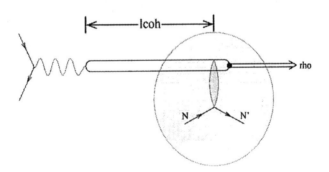

Figure 1. Cartoon describing exclusive ρ^0 meson production and illustrating the importance of coherence length, which is the propagation distance of the short-lived quark-antiquark state.

Fig. 2 presents data on the diffractive slope parameter b_p versus Q^2 for diffractive ρ^0 production from a proton target. This parameter characterizes the rate of exponential decay of the cross section with t, which is the squared four-momentum transfer between the vector meson and target nucleon. Physically, b is a measure of the transverse size of the interaction region. Fig. 2 demonstrates clearly that the virtual photon "shrinks" with increasing virtuality Q^2. When ρ^0 mesons are produced from a *nuclear* target rather than hydrogen, this shrinkage is one possible source of reduced final state interactions. Another arises from the coherence length $l_{coh} = \frac{2\nu}{Q^2+M^2_{q\bar{q}}}$, which describes the propagation distance of the short-lived quark-antiquark state (see Fig. 1). To separate the interplay of effects coming from the variation of transverse size and coherence length, we have performed a two-dimensional analysis of nuclear transparency as described in the Results section.

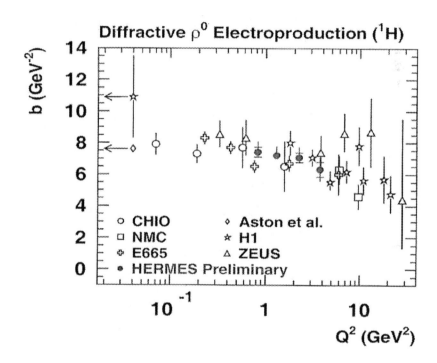

Figure 2. The measured Q^2-dependence of the diffractive slope parameter in exclusive ρ^0 production from a hydrogen target. The compilation of all results including HERMES data is taken from Ref. [3].

2. Event selection and kinematic reconstruction

The data were collected during the 1995-2000 running periods of the HERMES experiment using ^{1}H, ^{2}H, ^{3}He, ^{14}N, ^{20}Ne, and ^{84}Kr gas targets in the 27.5 GeV HERA positron storage ring at DESY. The HERMES detector is described in detail in Ref. [4]. The scattered e^+ and the h^+h^- hadron pair arising from the decay $\rho^0 \rightarrow \pi^+\pi^-$ were detected and identified in the HERMES forward spectrometer. The ρ^0 production sample was extracted from events with no more than these three detected tracks. A more detailed and comprehensive description of the extraction procedure for exclusive diffractive ρ^0 can be found in Ref. [5]. Here we only describe the method for extracting the ratio of coherent to incoherent cross sections based on nuclear data. Fig. 3 shows the exclusive $-t' = t - t_{min}$ distributions, where t is the four-momentum transfer described earlier and $|t_{min}|$ is the minimum $|t|$ allowed by the kinematics. The distributions exhibit the rapid falloff $\sim e^{-b|t'|}$ expected for diffractive processes and have two components: a coherent part, where the scattering occurs from the nucleus as a whole, and an incoherent part where the ρ meson scatters from a single quasifree nucleon within the target. The coherent component dominates at low $|t'|$ and is absent for the hydrogen target, while the incoherent part dominates starting at $|t'| \approx 0.05 GeV^2$.

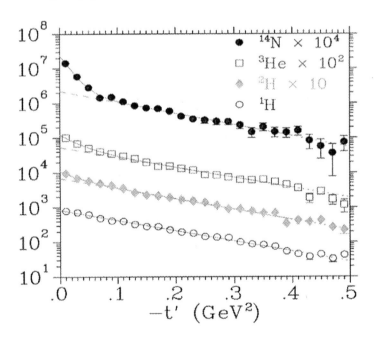

Figure 3. $|t'|$ spectra measured on different targets.

The cross-section is therefore approximated with the sum of coherent and incoherent contributions as:

$$\frac{d\sigma}{dt} = b_N e^{b_N t'} + R_A b_A e^{b_A t'}, \qquad (1)$$

where b_N and b_A are the slope parameters for the nucleon and nucleus respectively. This yields the first observable, the coherent to incoherent full cross-section ratio $R_A = \frac{\sigma_{coh}}{\sigma_{incoh}}$.

The coherent (incoherent) nuclear transparency is defined as follows:

$$Tr^{coh(incoh)} = \frac{\sigma^A_{coh(incoh)}(Q^2)}{A\sigma^p(Q^2)}, \qquad (2)$$

where σ^p refers to scattering from the proton, and A is the atomic number of the nuclear target. For the incoherent case nuclear transparency is associated with the probability that the produced ρ^0 meson escapes the nucleus without interaction. For the coherent case, measured for the first time at HERMES, such a probabilistic interpretation is not readily available, though the nuclear transparency is still sensitive to coherence length and color transparency effects.

3. Study of systematic effects

The systematic uncertainties for the measured ratios were determined by varying the event selection criteria or the analyses procedures. The sources of uncertainty include the finite vertex resolution of the tracking system and the background subtractions performed to determine ρ^0 mass spectrum and $-t'$ distribution. Ratios extracted in parallel for the $-t'$ and P_t^2 (where P_t^2 is transverse momentum square of the ρ^0 meson) spectra provided a measure of the extraction uncertainty. Monte-Carlo generator DIPSI [6] was used to calculate the detector acceptance. There different diffractive slope parameters and relativistic (nonrelativistic) Breit-Wigner mass distributions were used as an input parameters. Finally, corrections($\approx 15\%$) were applied to the ratios due to the "Pauli blocking" effect [7] for incoherent scattering. For the nuclear transparency measurement, the DIS positron cross-section was used as a luminosity measure in addition to the standard luminosity measurement based on Bhabha scattering from atomic electrons. The ratio of the integrated luminosities represents the largest source of kinematics-independent uncertainties. The total estimated systematic uncertainty from all normalization factors is 11%. The overall systematic uncertainty was determined by adding the above uncertainties in quadrature.

4. Results

In Fig. 4, the measured diffractive slope parameters for the different targets are presented along with a fit to their A-dependence. From earlier measurements of strong-interaction nuclear radii [8] an $A^{\frac{2}{3}}$ dependence is expected

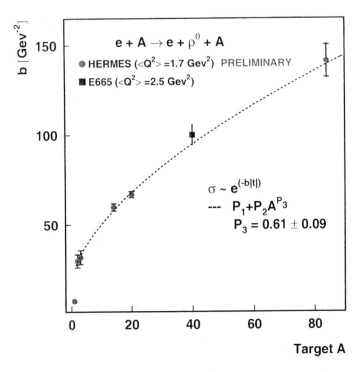

Figure 4. A dependence of the diffractive slope parameter b. The dashed line is the result of the fit. Also shown a point from E665 Fermilab Collaboration for calcium [11].

and is confirmed by the fit through the HERMES data.

Fig. 5 displays the coherent-to-incoherent cross section ratio R_A versus Q^2. A strong Q^2 dependence is observed, which is likely due to the variation of the coherence length l_{coh} and nuclear form factor Q^2 dependence. Fig. 7 presents the A-dependence of R_A. Note that it reaches saturation, which means that multiple scattering inside the nucleus does not grow linearly with A [7]. The data also indicate that the A-dependence of the coherent and incoherent mechanisms for exclusive ρ^0 electroproduction is similar. This observation is in contrast to the situation observed in J/ψ photo-production, where the coherent-to-incoherent cross section ratio was found to be A^α with $\alpha \approx 1.5$ [9]. In Fig. 6 the published HERMES data on incoherent transparency versus l_{coh} [5] are compared with the new coherent transparency result. As one can see, both the coherent and incoherent transparencies show a distinct l_{coh} dependence. To separate the color transparency effect, which is purely Q^2-dependent, from coherence length effects a two dimensional analysis was performed:

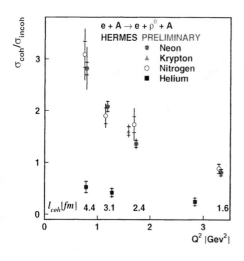

Figure 5. Q^2 dependence of the coherent to incoherent cross-section ratio. For each data point, the corresponding l_{coh} values are given.

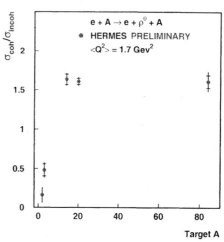

Figure 7. A dependence of the coherent to incoherent cross-section ratio.

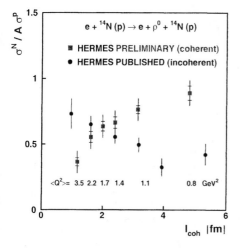

Figure 6. Nuclear Transparency for coherent and incoherent scattering on nitrogen target. For each data point, the coresponding Q^2 value is indicated.

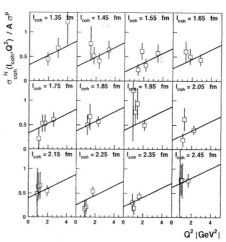

Figure 8. Nuclear transparency as a function of Q^2 in specific coherence length bins (as indicated in each panel) for coherent ρ^0 production on nitrogen. The straight line is the result of the *common* fit of the Q^2-dependence. The error bars include only statistical uncertainties.

the slope of the data with Q^2 was extracted at fixed l_{coh}. According to Ref. [10], a positive slope of the transparency with Q^2 is evidence for CT. Indeed, the results presented here support the CT prediction, as shown in Fig. 8 for coherent ρ^0 production. The results are also summarized in Table 1 for both coherent and incoherent ρ^0 production. When all data are combined together, the measured Q^2 slope is found to differ from zero by more than three standard deviations.

TABLE 1. Fitted common slope parameter of the Q^2 dependence of the nitrogen to hydrogen ratio with statistical and systematic uncertainties given separately. The results are compared to theoretical predictions [8]. The combined result represents the weighted mean of the coherent and incoherent ρ^0 data with uncertainties added in quadrature.

Data sample	Measured Q^2 slope (GeV^{-2})	Prediction (GeV^{-2})
coherent	$0.081 \pm 0.027 \pm 0.011$	0.060
incoherent	$0.097 \pm 0.048 \pm 0.008$	0.053
combined	0.085 ± 0.025(tot)	

References

1. T.H. Bauer et al., Rev. Mod. Phys. **50**, 261 (1978).
   H. Cheng and T. T. Wu, Phys. Rev. 183,1324(1969).
   J. D. Bjorken, J. B. Kogut, Phys. Rev. D 5,1152 (1972).
2. G. Bertsch et al., Phys. Rev. Lett. **47**, 297 (1981).
   A.B. Zamolodchikov et al., Pis'ma Zh. Eksp. Teor. Fiz. **33**, 612 (1981); Sov. Phys. JETP Lett. **33**, 595 (1981).
3. M. Tytgat, PhD Thesis, Gent University (2001); DESY-THESIS-2001-018.
4. HERMES Collaboration,K. Ackerstaff et al., Nucl. Instr. Meth. A**417**, 230 (1998).
5. HERMES Collaboration. K. Ackerstaff et al., Phys. Rev. Lett. **82**, 3025 (1999).
6. M. Arneodo et al., DESY96-149,(1996).
7. T. Renk, G. Piller, W. Weise, Nucl. Phys. A**689**, 869 (2001) ; J.S. Trefil, Nucl. Phys. B**11**, 330 (1969).
8. H. Alvensleben et al., Phys. Rev. Lett. **24**, 792 (1970).
9. E691 Collaboration, M.D. Sokoloff et al., Phys. Rev. Lett. **57**, 3003 (1986).
10. B.Z. Kopeliovich et al., Phys. Rev. C**65**, 035201(2002).
11. M.R. Adams et al., Phys. Rev. Lett. **74**, 1525 (1995).

ULTRAPERIPHERAL PHOTOPRODUCTION OF VECTOR MESONS IN THE NUCLEAR COULOMB FIELD AND THE SIZE OF NEUTRAL VECTOR MESONS

S.R. GEVORKYAN

Joint Institute for Nuclear Research,141980,Dubna,Russia
Yerevan Physics Institute, 375036, Yerevan, Armenia

I.P. IVANOV

IKP, Forschungszentrum Jülich,D-52425 Jülich, Germany
Sobolev Institute of Mathematics, 630090, Novosibirsk, Russia

AND

N.N. NIKOLAEV

IKP, Forschungszentrum Jülich,D-52425 Jülich, Germany
Landau Institute for Theoretical Physics, 142432,
Chernogolovka, Russia

Abstract. We point out a significance of ultraperipheral photoproduction of vector mesons in the Coulomb field of nuclei as a means of measuring the radius of the neutral vector meson. This new contribution to the production amplitude is very small compared to the conventional diffractive amplitude, but because of large impact parameters inherent to the ultraperipheral Coulomb mechanism its impact on the diffraction slope is substantial. We predict appreciable and strongly energy dependent increase of the diffraction slope towards very small momentum transfer.The magnitude of the effect is proportional to the mean radius squared of the vector meson and is within the reach of high precision photoproduction experiments, which gives a unique experimental handle on the size of vector mesons.

1. Introduction

It was noticed long time ago [1], that photoproduction of pseudoscalar mesons in the Coulomb field of nuclei allows one to measure the radiative width of mesons ($\pi^0 \to \gamma\gamma$, $V \to P\gamma$ etc., for the review and references see

E. Steffens and R. Shanidze (eds.), Spin Structure of the Nucleon, 243–251.

[2]). The principal point is that the Primakoff amplitude can be isolated for the presence of the Coulomb pole in the production amplitude, so that the contribution to the differential cross section of the process $\gamma + A \rightarrow P + A$, $P = (\pi^0, \eta, \eta\prime)$ from one-photon Coulomb exchange reads

$$\frac{d\sigma}{dq^2} = \frac{8\pi\alpha_{em}Z^2}{m_P^3}\Gamma_P\frac{q^2}{(q^2 + \Delta^2)^2}F_A^2(\vec{q})\,. \qquad (1)$$

Here m_P is the mass of the meson, Z is the nucleus charge number; $\alpha_{em} = \frac{e^2}{4\pi}$ is the fine structure constant, $q = |\vec{q}|$ and

$$\Delta = \frac{m_P^2}{2E_\gamma} \qquad (2)$$

are the transverse and the longitudinal momentum transfer, respectively, $F_A(q)$ is the nuclear charge form factor and Γ_P is the two-photon decay width of the pseudoscalar meson. The differential cross section (1) peaks at $q^2 = \Delta^2$, where it rises with the photon's energy E_γ as $\frac{d\sigma}{dt} \propto 1/\Delta^2 \propto E_\gamma^2$, whereas the position of the peak shifts to lower values of $q^2 \propto 1/E_\gamma^2$. This property allows one to isolate the Coulomb contribution unambiguously and measure the two-photon decay width $\Gamma_P = \Gamma(P \rightarrow 2\gamma)$ of pseudoscalar mesons. The recent progress of the experimental technique and high intensity photon beams available at CEBAF have lead to a new proposal of the measurement of the neutral pion lifetime at the level of several per mill, which would allow crucial tests of the chiral anomaly [3].

Subsequently, Pomeranchuk and Shmushkevich [4] extended the method to the determination of the lifetime of the Σ^0-hyperon via Coulomb production of the Σ^0 in the beam of Λ^0-hyperons. Subsequently, many radiative width of many charged meson resonances have been measured via Coulomb photoproduction of resonances in the pion and kaon beams (for the review and references see [2] and the Review of Particle Properties [5]). Because of the C-parity constraints, the Primakoff effect does not contribute to the photoproduction of the vector mesons, and one needs at least two-photon exchange. The two-photon exchange contribution to the photoproduction amplitude will no longer contain the Coulomb pole inherent to the Primakoff contribution. However, the principal point that in the Coulomb amplitude the important impact parameters are very large,

$$|\vec{b}| \sim \frac{1}{\Delta} \qquad (3)$$

remains very much relevant. Furthermore, according to (2), the range of relevant impact parameters rises rapidly with energy. The Coulomb contribution to the photoproduction amplitude from these large impact parameters has logarithmic singularity $\propto \log[1/R_A^2(\vec{q}^2 + \Delta^2)]$, which generates the singular dependence of the forward diffraction slope on q^2 and energy.Finally, invoking the familiar vector meson dominance, the Coulomb contribution to the photoproduction amplitude can be related to the amplitude of elastic scattering of the vector meson in the Coulomb field. To the two-photon exchange approximation, the latter is proportional to the radius squared of the vector meson. Consequently, the experimental isolation of the ultraperipheral Coulomb contribution to the diffractive vector meson production would lead to a unique experimental measurement of the size of neutral vector mesons and must not be overlooked.

The estimations of higher order Coulomb corrections have been performed earlier for photoproduction of pseudoscalar mesons [6]. There, too, multiple Coulomb exchanges have the logarithmic singularity, but they are too weak to produce a numerically substantial correction to the Primakoff amplitude with the pole singularity.

2. Color and electric dipole view at the ultraperipheral Coulomb production and the radius of vector mesons

Because of the rise of the coherence and formation times, at high energies the meson photoproduction can be viewed as a three step process: splitting of the photon into quark-antiquark pair at a large distance in front of the target, interaction of the quark-antiquark color and electric dipole with the target, and projection of the scattered quark-antiquark pair onto the observed meson. When such a target is a Coulomb field of heavy nuclei, it is not á priori obvious that higher order Coulomb corrections will be small, because the QED expansion parameter $Z\alpha_{em} \sim 1$. However, because mesons are electrically neutral and have a small size R_M, the strength of the interaction of small electric dipole at large impact parameters $\vec{b}$ is suppressed by the small parameter $R_M^2/\vec{b}^2$, so that the ultraperipheral Coulomb production can be estimated to the leading two-photon exchange approximation.

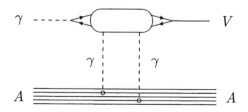

Figure 1. The two-photon exchange contribution to the photoproduction amplitude

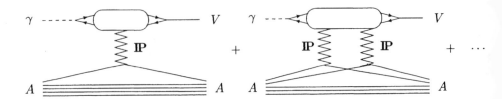

Figure 2. Left: the single-pomeron exchange, i.e., impulse approximation, contribution to the photoproduction off a nucleus,right: the multiple exchange contribution to the the photoproduction off a nucleus.

The technical argument why the Coulomb contribution is singular and measures the size of the vector meson goes as follows. The vector meson photoproduction amplitude $\gamma + A \rightarrow V + A$ is the sum of Coulomb (Fig. 1), M_C, and strong (Fig. 2), M_s, production amplitudes. As we argued, the Coulomb amplitude can be evaluated to the lowest order in QED perturbation theory, i.e. the two-photon exchange approximation. Because the typical impact parameters which contribute to the Coulomb amplitude are much larger than the size of the vector meson, $|\vec{b}| \sim 1/\Delta = 2E_\gamma/m_V^2 \gg R_V$, the principal quantity is an imaginary part of an amplitude, $f(\vec{r}, \vec{q})$, of a nearly forward scattering of the $q\bar{q}$ electric dipole in the Coulomb field of a nucleus. With some important modifications for the effect of the finite longitudinal momentum transfer, which amounts to an effective screening of the Coulomb potential, the amplitude $f(\vec{r}, \vec{q})$ is related to the electric dipole scattering cross section $\sigma_C(\vec{r})$, where $\vec{r}$ is the $q\bar{q}$ separation in the two-dimensional impact parameter space. The latter enters, for instance, the calculation of a scattering of atoms in a Coulomb field, the case of muonium (or pionium [7]) is the most relevant one because the size of muonium is much smaller than the radius of an atom and nucleus acts predominantly as a pointlike charge. If we define $f(\vec{r}, \vec{q})$ for the electric dipole made of the particles of charge ± 1, then the imaginary part of the Coulomb amplitude can be cast in the form

$$M_C(\vec{q}) = C_C(V)\langle V|f(\vec{r}, \vec{q})|\gamma > = \frac{e}{F_V}\langle V|f(\vec{r}, \vec{q})|V\rangle \tag{4}$$

where the charge-isotopic factors equal

$$C_C(\rho) = \frac{1}{\sqrt{2}}(e_u^3 + e_d^3) = \frac{1}{3\sqrt{2}}$$
$$C_C(\omega) = \frac{1}{\sqrt{2}}(e_u^3 - e_d^3) = \frac{7}{27\sqrt{2}}$$
$$C_C(\phi) = e_s^3 = \frac{1}{27} \tag{5}$$

In (4) we invoked vector dominance model (for an extensive review and references see [8]) and F_V are related to the standard vector dominance constants f_V as

$$\frac{1}{F_\rho} = \frac{1}{f_V}(e_u^2 - e_u e_d + e_d^2) = \frac{7}{9f_\rho}$$
$$\frac{1}{F_\omega} = \frac{1}{f_V}(e_u^2 + e_u e_d + e_d^2) = \frac{1}{3f_\omega}$$
$$\frac{1}{F_\phi} = \frac{1}{f_\phi}e_s^2 = \frac{1}{9f_\phi} \qquad (6)$$

The amplitude (4) is normalized so that

$$\frac{d\sigma}{dq^2} = \frac{|M|^2}{16\pi}. \qquad (7)$$

In the calculation of the amplitude $f(\vec{r}, \vec{q})$ a care must be taken of the longitudinal momentum transfer. Interaction of the incoming photon with the first nuclear photon puts the quark-antiquark pair on mass shell, and the exchanged photon has a longitudinal momentum

$$\Delta_1 = \frac{M^2}{2E_\gamma}, \qquad (8)$$

where M is the invariant mass of the $q\bar{q}$ pair. The second photon has a longitudinal momentum

$$\Delta_2 = \frac{M_V^2 - M^2}{2E_\gamma}. \qquad (9)$$

In the nonrelativistic quark model $M \sim M_V$ and $|\Delta_2| \ll \Delta_1$. This hierarchy is a good starting approximation for light mesons as well. In what follows we put $\Delta_1 = \Delta = m_V^2/2E_\gamma$ and $\Delta_2 = 0$.

The relevant two-photon exchanged dipole amplitude (Fig.1) can be cast in the form:

$$f(\vec{r}, \vec{q}) = 8(Z\alpha_{em})^2 \int \frac{d^2k(e^{\frac{i\vec{q}\vec{r}}{2}} - e^{i\vec{k}\vec{r}})F_A(\vec{k} - \frac{1}{2}\vec{q})F_A(\vec{k} + \frac{1}{2}\vec{q})}{[(\vec{k} - \frac{1}{2}\vec{q})^2 + \Delta^2](\vec{k} + \frac{1}{2}\vec{q})^2}, \qquad (10)$$

For high energy photons the size of the vector meson can be neglected compared to $1/\Delta$. In addition we are interested in the very forward production $q^2 \sim \Delta^2$. Then one can safely expand the exponentials and neglect q in the nuclear form factors, which allows to do integration in (10) with the following result

$$f(\vec{r}, \vec{q}) = 2\pi (Z\alpha_{em})^2 r^2 \ln \frac{6}{(q^2 + \Delta^2)\langle R_A^2 \rangle_{ch}}, \qquad (11)$$

which clearly exhibits the logarithmic singularity of the Coulomb amplitude as a function of q^2. Here $\langle R_A^2 \rangle_{ch}$ is the charge radius of the nucleus squared, and hereafter the capital $\vec{R}$ stands for the 3-dimensional radius-vector. The substitution of (11) into (4) gives

$$\begin{aligned} M_C(\vec{q}) &= \frac{2\pi e(Z\alpha_{em})^2}{F_V} \langle V | \vec{r}^2 | V \rangle \ln \frac{6}{(q^2 + \Delta^2)\langle R_A^2 \rangle_{ch}} \\ &= \frac{16\pi e(Z\alpha_{em})^2}{3F_V} \langle \vec{R}_V^2 \rangle_{ch} \ln \frac{6}{(q^2 + \Delta^2)\langle R_A^2 \rangle_{ch}} \qquad (12) \end{aligned}$$

where in the case of the $\rho^{\pm}$-meson $\langle \vec{R}_V^2 \rangle_{ch}$ is the conventionally defined 3-dimensional charge radius mean squared. Consequently, the experimental isolation of the ultraperipheral Coulomb production amplitude would give a unique possibility to measure the radius of the vector meson squared.

3. The amplitude of photoproduction via strong interaction

Within the QCD color dipole framework, one needs first to calculate the amplitude of interaction of the $q\bar{q}$ color dipole with the target, and then use the same formalism as outlined above for electric dipoles [9]. However, for the illustration purposes it is more straightforward to invoke the vector dominance model by which

$$M_s(\vec{q}) = \frac{e}{f_V} M_{VA}(\vec{q}). \qquad (13)$$

Anyway, the real photoproduction of vector mesons is a soft process and $M_s(\vec{q})$ can not be reliably calculated in the pQCD framework. The vector meson-nucleus scattering amplitude can be evaluated in the standard Glauber approximation. For a simple as yet sufficiently accurate Gaussian parameterization for a nuclear matter density,

$$n_A(r) = n_0 \exp\left(-\frac{r^2}{R_A^2}\right), \qquad (14)$$

one finds the familiar expression

$$M_{VA}(\vec{q}) = A\sigma_{VN} \exp\left(\frac{R_A^2 q^2}{4A}\right) \sum_{n=1}^{A} \frac{(A-1)!}{n \cdot n!(A-n)!}$$

$$\times \left[-\frac{\sigma(VN)}{2\pi(R_A^2 + 2B_N)} \right]^{n-1} \exp\left(-\frac{(R_A^2 + 2B_N)q^2}{4n} \right), \quad (15)$$

where B_N is the diffraction slope of the VN elastic scattering. In (15) we neglected the small real part of the VN scattering amplitude. The total cross section of the vector meson interaction with nucleon $\sigma(VN)$ can be obtained from the relation

$$\sigma(VN) = \frac{\sigma(\pi^+ N) + \sigma(\pi^- N)}{2}.$$

For the estimation purposes, one can neglect B_N compared to R_A^2 and use the simple parameterization

$$M_{VA}(\vec{q}) \approx \sigma(VN) A^\alpha \exp\left[-\frac{1}{2} B_A \vec{q}^{\,2} \right] \quad (16)$$

where for the ρ and ω the exponent $\alpha \approx 0.8$ and to the crude approximation $B_A \approx \frac{1}{3}\langle R_A^2 \rangle_{ch}$.

4. Is the Coulomb correction observable?

The Coulomb correction to the differential cross section is small,

$$\frac{2M_C}{M_s} = \eta_C \ln \frac{6}{(\vec{q}^{\,2} + \Delta^2)\langle \vec{R}_A^2 \rangle_{ch}}$$

$$= \frac{32\pi(Z\alpha_{em})^2 f_V}{3 F_V A^\alpha} \cdot \frac{\langle \vec{R}_V^2 \rangle_{ch}}{\sigma(VM)} \cdot \ln \frac{6}{(\vec{q}^{\,2} + \Delta^2)\langle \vec{R}_A^2 \rangle_{ch}}. \quad (17)$$

Take the photoproduction of the ρ meson off the lead target, for which $\langle \vec{R}_A^2 \rangle_{ch} \approx 30 f^2$. The nonrelativistic quark model suggests [10] $\langle \vec{R}_V^2 \rangle_{ch} \approx \langle \vec{R}_\pi^2 \rangle_{ch} \approx 0.439 f^2$ and $\sigma(\rho N) \sim 30mb$. For $E_\gamma = 10 GeV$ the logarithmic factor is of the order of 2, so that the Coulomb correction is small $\sim 4\%$.

The logarithmic singularity of $2M_C(\vec{q})$ makes the Coulomb correction much more noticeable in the diffraction slope

$$B = 2\frac{d\ln M}{d\vec{q}^{\,2}} \quad (18)$$

Using the above results one obtains for the slope of the full amplitude

$$B = B_A + \Delta B_C = B_A + \eta_C \frac{1}{\vec{q}^{\,2} + \Delta^2} = B_A + \frac{4\eta_C E_\gamma^2}{m_\rho^4} \cdot \frac{\Delta^2}{\vec{q}^{\,2} + \Delta^2}. \quad (19)$$

In order to emphasize the importance of the Coulomb correction to the diffraction slope, ΔB_C, start with the exactly forward ($\vec{q}^2 = 0$) photoproduction off the lead target. Here the Coulomb correction rises from $\Delta B_C \sim 25 GeV^{-2}$ at $E_\gamma = 10$ GeV to a dramatically large $\Delta B_C \sim 2500$ GeV^{-2} at $E_\gamma = 100$ Gev to be compared to $B_A \sim 250 GeV^{-2}$. Of course, this dramatic rise is confined to a region of ultrasmall momentum transfers, from $|\vec{q}| \sim \Delta \approx 2.5$MeV/c at $E_\gamma = 100$ GeV, which are perhaps beyond the experimental reach, to a more realistic $|\vec{q}| \sim \Delta \approx 25$MeV/c at $E_\gamma = 10$ GeV. Now notice, that while the prefactor η_C does not depend on our approximations for $\Delta_{1,2}$, the exact $\vec{q}^2$ dependence, and the magnitude at $\vec{q}^2 = 0$, of the ΔB_C, do explicitly depend on $\Delta_{1,2}$. Consequently, for the accurate determination of the size of the ρ-meson one must concentrate on the Coulomb correction ΔB_C at $\vec{q}^2 > \Delta^2$, where the $\vec{q}^2$ dependence is insensitive to $\Delta_{1,2}$.

5. Summary and conclusions

We pointed out a new, ultraperipheral Coulomb, mechanism for the photoproduction of vector mesons off nuclei at ultrasmall momentum transfer. The Coulomb contribution to the photoproduction amplitude has a singular, logarithmic, dependence on the momentum transfer squared, and is proportional to the mean radius of the vector meson squared. This strongly energy dependent singular contribution entails a steep dependence of the diffraction slope on $\vec{q}^2$. The very specific dependence of the Coulomb correction ΔB_C on the photon energy, the momentum transfer squared $\vec{q}^2$, the target charge and mass number makes the experimental evaluation of the size of the ρ meson quite plausible. The work of I.I. and S.G. has been partly supported by the grants INTAS 00-0036 and INTAS 97-30494.

References

1. Primakoff, H.(1951) Photoproduction of neutral mesons in nuclear electric fields and the mean life of the neutral mesons, Phys. Rev. 81, 899
2. Landsberg, L.G.(1985) Electromagnetic decays of light mesons, Phys. Reports 128, 303
3. Gasparian, A. et al.(1998) A precision measurement of the neutral pion lifetime via the Primakoff effect, TJNAF Proposal PR-98-105
4. Pomeranchuk, I.Y.and Shmushkevich, I.M.(1961) On processes in the interaction of γ-quanta with unstable particles, Nucl. Phys. 23, 452
5. Review of Particle Properties (1998) The European Phys.J., 3, 1
6. Schwiete, G. (2000) The higher order corrections to Primakoff effect, Acta Phys. Polon. B31, 2437 ; Diploma Thesis, Institut f. Kernphysik, Forschungszentrum Jülich & University of Bonn, 2000.
7. Gevorkyan, S.R., Tarasov, A.V. and Voskresenskaya, O.O.(1998) Total Cross Sections for the Interaction of Hydrogen-like Atoms with Atoms of a Medium, Physics of Atomic Nuclei 61, 1507

8. Bauer, T.H., Spital, R.D., Yennie, D.R. and Pipkin, F.M.(1978) The hadronic properties of the photon in high energy interactions, Rev. Mod. Phys. 50, 261
9. Nemchik,J., Nikolaev,N.N. and Zakharov,B.G. (1994) Scanning the BFKL pomeron in elastic production of vector mesons at HERA, Phys. Lett. B341, 228
10. Armendolia S.R. et al. (1986) A measurement of the space-like pion electromagnetic form factors,Nucl.Phys. B277,168

CROSSING SYMMETRY VIOLATION IN THE PROCESS OF LEPTON PAIR PRODUCTION IN RELATIVISTIC ION COLLISIONS COMPARED WITH THE CROSSING PROCESS

E. A. KURAEV, E. BARTOŠ AND S. R. GEVORKYAN

Joint Institute of Nuclear Research, 141980 Dubna, Russia

Abstract. Using the Sudakov technique we sum the perturbation series for the process $3 \to 3$ and obtain the compact analytical expression for the amplitude of this process, which takes into account all possible Coulomb interactions between colliding particles. Compare it with the amplitude of the lepton pair production in heavy ion collision, i.e., in the process $2 \to 4$, we show that crossing symmetry between this processes holds only if one neglects the interaction of produced pair with ions (i.e. in the approximation $Z_{1,2}\alpha \ll 1$).

1. Introduction

During the past decade the growing interest to the process of lepton pair production in the strong Coulomb fields appeared. This is connected mainly with beginning of the operation of the relativistic heavy ion collider RHIC (Lorentz factor $\gamma = \frac{E}{m} = 108$) and the new collider LHC ($\gamma = 3000$) which will operate in the nearest future. At such energies the lepton pair yield becomes huge (according to [1, 2]), so that a detail analysis of the process

$$A + B \to A + B + e^+ + e^- \tag{1}$$

accounting the Coulomb corrections (CC) is required. Such work has been done during last years and a lot of papers are devoted to this subject [3, 4, 5, 6, 7, 8, 11, 12, 13]. Nevertheless, the problem turn out to be more complex than it seems from the first glance. We want only to notice the exciting result obtained in the papers [3, 4, 7]: the Coulomb corrections to the process (1) enter the amplitude of this process in such a way that its cross

E. Steffens and R. Shanidze (eds.), Spin Structure of the Nucleon, 253–263.

section is determined solely by the lowest (Born) term. At present we understand that this result is the incorrect application of crossing symmetry property which, as it is known long ago (see, e.g., [14]), is valid only on the Born level. As an obvious example of the crossing symmetry violation we want to cite the process of lepton pair photoproduction on the nuclei and its counterpart, the bremsstrahlung in lepton–nucleus scattering. Amplitudes of both processes are determined by Coulomb phase which is infrared stable in the case of pair photoproduction whereas it is infrared divergent in the case of bremsstrahlung and this difference cannot be adjust by trivial crossing change of variables. Taking into account the importance of the problem and permanent interest to it from the scientific society, we calculated the full amplitude for the process (2) accounting all possible photon exchanges among the colliding relativistic particles. Comparing it with the amplitude of the process (1) we had shown that the crossing symmetry property becomes invalid whereas one takes into account the final state interaction of the lepton pair with the Coulomb field of ions.

2. The Born amplitude of the process $3 \to 3$

Let us construct the amplitude of the process $3 \to 3$ represented in Fig. 1 (a, b)

$$A_1(p_1) + A_2(p_2) + C(p_3) \to A_1(p_1') + A_2(p_2') + C(p_3'). \tag{2}$$

We consider the kinematics when all the energy invariants which determined the process (2) are large, compared with the masses of involved particles and the transfer momenta

$$s = (p_1 + p_2)^2, \quad s_1 = (p_1 + p_3)^2, \quad s_2 = (p_3 + p_2)^2,$$
$$q_1^2 = (p_1 - p_1')^2, \quad q_2^2 = (p_2 - p_2')^2, \quad q_3^2 = (p_3 - p_3')^2, \tag{3}$$
$$p_1^2 = {p_1'}^2 = m_1^2, \quad p_2^2 = {p_2'}^2 = m_2^2, \quad p_3^2 = {p_3'}^2 = m^2,$$
$$s \gg s_1 \sim s_2 \gg -q_1^2 \sim -q_2^2 \sim -q_3^2.$$

For the Born amplitude of the process (2) one can write

$$M_{(1)}^{(1)} = -i(4\pi\alpha)^2 Z_1 Z_2 \bar{u}(p_1')\gamma_\mu u(p_1)\bar{u}(p_2')\gamma_\nu u(p_2)\frac{\bar{u}(p_3')O_{\rho\sigma}u(p_3)g^{\mu\sigma}g^{\nu\rho}}{q_1^2 q_2^2}, \tag{4}$$

where $Z_{1,2}$ are the charge numbers of the colliding nuclei. We use Sudakov parameterization for all four–momenta entering the problem (for details see

[12])

$$q_1 = \alpha_1 \tilde{p}_2 + \beta_1 \tilde{p}_1 + q_{1\perp}, \quad q_2 = \alpha_2 \tilde{p}_2 + \beta_2 \tilde{p}_1 + q_{2\perp},$$
$$p_1' = \alpha_1' \tilde{p}_2 + \beta_1' \tilde{p}_1 + p_{1\perp}', \quad p_2' = \alpha_2' \tilde{p}_1 + \beta_2' \tilde{p}_1 + p_{2\perp}', \tag{5}$$
$$p_3 = \alpha_3 \tilde{p}_1 + \beta_3 \tilde{p}_1 + p_{3\perp}, \quad p_3' = \alpha_3' \tilde{p}_1 + \beta_3' \tilde{p}_1 + p_{3\perp}',$$

and the Gribov decomposition of the metric tensor into the longitudinal and transverse parts

$$g_{\mu\nu} = g_{\perp\mu\nu} + \frac{2}{s}\left(\tilde{p}_{1\mu}\tilde{p}_{2\nu} + \tilde{p}_{1\nu}\tilde{p}_{2\mu}\right),$$

with light–like 4-vectors $\tilde{p}_{1,2}$. For the kinematics of the process we will use the following relations

$$s = 2\tilde{p}_1\tilde{p}_2, \quad \beta_1 + \beta_3 = \beta_3', \quad \alpha_2 + \alpha_3 = \alpha_3', \tag{6}$$
$$g_{\mu\sigma} = \frac{2}{s}\tilde{p}_{1\sigma}\tilde{p}_{2\mu}, \quad g_{\nu\rho} = \frac{2}{s}\tilde{p}_{1\nu}\tilde{p}_{2\rho},$$
$$q_1^2 = q_{1\perp}^2 = -\mathbf{q_1}^2, \quad q_2^2 = q_{2\perp}^2 = -\mathbf{q_2}^2,$$

where $\mathbf{q_i}$ are two–dimensional vectors in the plane transverse to the z-axes, which we choose along 3–vector $\vec{p}_1 = -\vec{p}_2$ in the center of mass frame of initial particles A_1, A_2. Using the gauge invariant condition

$$q_{1\rho}\bar{u}(p_3')O_{\rho\sigma}u(p_3) \approx (\beta_1\tilde{p}_1 + q_{1\perp})_\rho \bar{u}(p_3')O_{\rho\sigma}u(p_3) = 0,$$
$$q_{2\sigma}\bar{u}(p_3')O_{\rho\sigma}u(p_3) \approx (\alpha_2\tilde{p}_2 + q_{2\perp})_\sigma \bar{u}(p_3')O_{\rho\sigma}u(p_3) = 0, \tag{7}$$

one gets the Born amplitude in the form

$$M_{(1)}^{(1)}(q_1, q_2) = -4is N_1 N_2 (4\pi\alpha Z_1)(4\pi\alpha Z_2) B(q_1, q_2), \tag{8}$$

with

$$B(q_1, q_2) = \frac{\bar{u}(p_3')O_{\rho\sigma}u(p_3)q_{1\perp}^\rho q_{2\perp}^\sigma}{s\alpha_2\beta_1\mathbf{q_1}^2\mathbf{q_2}^2}, \tag{9}$$
$$N_1 = \frac{1}{s}\bar{u}(p_1')\hat{p}_2 u(p_1), \quad N_2 = \frac{1}{s}\bar{u}(p_2')\hat{p}_1 u(p_2),$$
$$s\alpha_2\beta_1 = -q_3^2 - (\mathbf{q_1} - \mathbf{q_2})^2 \sim \mathbf{m}^2.$$

The values of N_i for every polarization state of initial particles (or for spinless particles) are unity and

$$\bar{u}(p_3')O_{\rho\sigma}u(p_3)q_{1\perp}^\rho q_{2\perp}^\sigma =$$
$$\bar{u}(p_3')\left[\hat{q}_{2\perp}\frac{\hat{p}_3 + \hat{q}_1 + m}{(p_3 + q_1)^2 - m^2}\hat{q}_{1\perp} + \hat{q}_{1\perp}\frac{\hat{p}_3 + \hat{q}_2 + m}{(p_3 + q_2)^2 - m^2}\hat{q}_{2\perp}\right]u(p_3). \tag{10}$$

3. The Coulomb corrections to the process $3 \to 3$

Let us consider the set of six Feynman diagrams (FD) with one virtual photon connected the p_3 line with the particle A_1 and the two ones connected p_3 line with the particle A_2 (see Fig. 2). The loop momentum integration in the relevant matrix element can be performed accounting that

$$d^4k = (2\pi i)^2 \frac{1}{2s} \frac{d(s\alpha_k)}{2\pi i} \frac{d(s\beta_k)}{2\pi i} d^2k_\perp, \quad k = \alpha_k \tilde{p}_2 + \beta_k \tilde{p}_1 + k_\perp. \quad (11)$$

It can be shown that only 4 FD amplitudes works (Fig. 2 (a–d)). Really, when one write explicitly denominators in Fig. 2 (e, f) through longitudinal Sudakov variables, i.e.,

$$
\begin{aligned}
(p_3 + k)^2 - m^2 + i0 &\approx s\alpha_k \beta_3 + i0, \\
(p_3 - q_2 + k)^2 - m^2 + i0 &\approx s\alpha_k \beta_3 + i0, \\
(p_2 - k)^2 - m^2 + i0 &\approx -s\beta_k + i0, \\
(p_2' + k)^2 - m^2 + i0 &\approx s\beta_k + i0,
\end{aligned}
\quad (12)
$$

one can see that both poles in α_k complex plane are situated in the same half-plane, so their contribution to the amplitude is zero (suppressed by factor $|q_3^2/s| \sim |s_1/s|$). This result is in agreement with one obtained in [8].

It is convenient to introduce 8 FD (including four ones depicted in Fig. 2 (a–d) and additional four FD with interchanged photons absorbed by nucleus A_2 line). To avoid the double counting we multiply the relevant matrix element by statistical factor $1/2!$. This trick permits one to perform the integration over α_k, β_k with the result

$$\int\limits_{-\infty}^{\infty} \frac{d(s\alpha_k)}{2\pi i} \left(\frac{\beta_3}{s\alpha_k \beta_3 + i0} + \frac{\beta_3}{-s\alpha_k \beta_3 + i0} \right) =$$

$$\int\limits_{-\infty}^{\infty} \frac{d(s\beta_k)}{2\pi i} \left(\frac{1}{s\beta_k + i0} + \frac{1}{-s\beta_k + i0} \right) = 1. \quad (13)$$

Now let us show how the cancellations of contribution arising from FD with absorbtion of $n+1$ number of exchanged photons between particle C and nucleus A_1, sandwiched between two exchanges between particle C and nucleus A_2 (Fig. 3) take place. The algebraic symmetrization procedure described above (13), with using the relations

$$l_i = \alpha_{l_i} \tilde{p}_2 + \beta_{l_i} \tilde{p}_1 + l_{i\perp}, \quad \alpha_{l_i} < \alpha_3 < \alpha_k \le 1, \quad \beta_k < \beta_3 < \beta_{l_i} \le 1, \quad (14)$$

leads to the product of factors

$$\prod_{i=1}^{n} \left(\frac{\beta_3}{s\alpha_{l_i}\beta_3 + i0} + \frac{\beta_3}{-s\alpha_{l_i}\beta_3 + i0} \right) \prod_{j=1}^{n} \left(\frac{1}{s\beta_{l_j} + i0} + \frac{1}{-s\beta_{l_j} + i0} \right), \quad n \geq 1.$$
(15)

In the terms of notation used in [9, 10] our assumptions (14) read

$$E = p_{1+} \gg p_{3+} \gg p_{2+} = \frac{m_2^2}{E}, \quad \frac{m_1^2}{E} = p_{1-} \ll p_{3-} \ll p_{2-} = E. \quad (16)$$

It is easy to see, that in this case, no dependence on $p_{3\pm}$ sign appears for the eikonal amplitudes corresponding to the situation in Fig. 3.

The poles of the electron Green functions (Fig. 3) are situated at the same half plane of k_- what one allows to safely neglect the contribution of such diagrams.

Performing the integration over longitudinal Sudakov variables α_{l_i}, β_{l_j} in blocks 1, 2 of FD in Fig. 3 one can see, that dependence on longitudinal Sudakov variables α_k, β_k relevant to the lower loop is completely the same as in the previous case (see Fig. 2 (e, f)), therefore contribution of such type of FD to the total amplitude is zero.

The physical reason of this suppression is the same as in the case of bremsstrahlung suppression for fast charged particle moving through the media known as the Landau–Pomeranchuk–Migdal effect. Really, this effect can be explained starting from the fact of power suppression of radiation between two scattering centers in the case when the distance between these centers is less than the coherence length.

Further integration over transverse momentum is straightforward

$$\int \frac{d^2\mathbf{k}}{\pi} \frac{1}{(\mathbf{k} + \lambda)((\mathbf{q_2} - \mathbf{k})^2 + \lambda^2)} = \frac{2}{\mathbf{q_2}^2} \ln \frac{\mathbf{q_2}^2}{\lambda^2}. \quad (17)$$

For the amplitude $M_{(2)}^{(1)}$ (see Fig. 4) and the similar amplitude $M_{(1)}^{(2)}$ we obtained

$$M_{(2)}^{(1)} + M_{(1)}^{(2)} = M_{(1)}^{(1)}(q_1, q_2) \frac{1}{2!} \left[2iZ_1\alpha \ln \frac{\mathbf{q_1}^2}{\lambda^2} + 2iZ_2\alpha \ln \frac{\mathbf{q_2}^2}{\lambda^2} \right]. \quad (18)$$

The amplitude for arbitrary amount of interchanged photons (see Fig. 5) is constructed in the similar way

$$M_{(\infty)}^{(\infty)}(q_1, q_2) = M_{(1)}^{(1)}(q_1, q_2)e^{i(\varphi_1(\mathbf{q_1}) + \varphi_2(\mathbf{q_2}))} \quad (19)$$

with the Coulomb phases

$$\varphi_1(\mathbf{q_1}) = \mathbf{Z_1}\alpha \ln \frac{\mathbf{q_1}^2}{\lambda^2}, \quad \varphi_2(\mathbf{q_2}) = \mathbf{Z_2}\alpha \ln \frac{\mathbf{q_2}^2}{\lambda^2}.$$

Consider now the case with one additional exchanged photon between two nuclei A_1, A_2. The relevant matrix element $M_{(1B)}$ reads

$$M_{(1B)} = i\alpha Z_1 Z_2 \int \frac{d^2\mathbf{k}}{\pi(\mathbf{k}^2 + \lambda^2)} M_{(1)}^{(1)}(q_1 + k, q_2 + k). \qquad (20)$$

Two-dimensional integral in (20) is infrared divergent. To regularize it we introduce the photon mass parameter λ.

In the same approach we get for the matrix element with the n exchanged (between nuclei) photons (see Fig. 6 (a))

$$M_{(nB)} = \frac{(i\alpha Z_1 Z_2)^n}{n!} \prod_{i=1}^{n} \int \frac{d^2\mathbf{k_i}}{\pi(\mathbf{k_i}^2 + \lambda^2)} M_{(1)}^{(1)}(q_1 + \sum_{i=1}^{n} k_i, q_2 + \sum_{i=1}^{n} k_i). \qquad (21)$$

It is convenient to write down this expression in the impact parameter representation. For this aim we use the following identity

$$\int d^2\mathbf{k_{n+1}}\delta^{(2)}(\mathbf{k_{n+1}} - \mathbf{q_1} - \sum_{i=1}^{n} \mathbf{k_i}) =$$

$$\frac{1}{(2\pi)^2} \int d^2\mathbf{k_{n+1}} d^2\rho e^{i(\mathbf{k_{n+1}} - \mathbf{q_1} - \sum \mathbf{k_1})\cdot\rho} = 1. \qquad (22)$$

Thus the matrix element with arbitrary number of exchanged photons can be cast

$$M(3 \to 3) = \sum_{n=1}^{\infty} M_{(nB)} = \frac{1}{4} \int \frac{d^2\rho}{\pi} e^{-iq_1 \cdot \rho} e^{i\alpha Z_1 Z_2 \psi(\rho)} \tilde{M}_{(1)}^{(1)}(\rho, q_1, q_2) \qquad (23)$$

with

$$\psi(\rho) = \int \frac{d^2\mathbf{k}}{\pi} \frac{e^{-i\mathbf{k}.\rho}}{\mathbf{k}^2 + \lambda^2} = 2K_0(\rho\lambda) \approx -2\ln\left(\frac{C\rho\lambda}{2}\right), \qquad (24)$$

where $C \approx 1.781$ and

$$\tilde{M}_{(1)}^{(1)}(\rho, q_1, q_2) = \int \frac{d^2\mathbf{k}}{\pi} e^{-i\mathbf{k}.\rho} M_{(1)}^{(1)}(k, k + q_2 - q_1). \qquad (25)$$

This result confirms the general ansatz given above (see (19)) that the dependence on "photon mass" λ can be represented as a phase factor. As can be seen the whole amplitude (23) cannot be cast solely as a Born amplitude multiplied by the phase factor. The corresponding contributions to the total cross section (except the Born term) will be enhanced only by the first power of logarithm in energy.

Finally, taking into account all photon exchanges between particle C and nuclei A_1, A_2 we obtain the general answer by the simple replacement in the expression (25)

$$M_{(1)}^{(1)} \to M_{(\infty)}^{(\infty)} = M_{(1)}^{(1)}(k, k + q_2 - q_1)e^{i\varphi_1(\mathbf{k})+i\varphi_2(\mathbf{k}+\mathbf{q_2}-\mathbf{q_1})} \qquad (26)$$

with φ_1, φ_2 given in (19).

4. The Coulomb corrections to the process of lepton pair production

As was mentioned above, our goal is to investigate the crossing symmetry property between the amplitudes of the process (2) and the relevant process in Fig. 1 (c, d)

$$A_1(p_1) + A_2(p_2) \to A_1(p_1') + A_2(p_2') + C(p_3) + \bar{C}(p_4), \qquad (27)$$

with the following kinematics

$$s = (p_1 + p_2)^2, \quad s_p = (q_+ + q_-)^2, \quad q_1^2 = (p_1 - p_1')^2, \quad q_2^2 = (p_2 - p_2')^2, \qquad (28)$$

$$p_1^2 = p_1'^2 = m_1^2, \quad q_+^2 = q_-^2 = m_2^2, \quad q_+^2 = q_-^2 = m^2,$$

$$s \gg -q_1^2 \sim -q_2^2 \sim s_{12}.$$

Using the Sudakov technique the Born amplitude for the process (27) can be represented in the form

$$M_p = -is2^6\pi^2 Z_1 Z_2 N_1 N_2 B_p(q_1, q_2) \qquad (29)$$

with

$$B_p(q_1, q_2) = \frac{e_1^\alpha e_2^\beta \bar{u}(q_-)T_{\alpha\beta}v(q_+)|\mathbf{q_1}||\mathbf{q_2}|}{\tilde{s}\mathbf{q_1}^2\mathbf{q_2}^2},$$

$$\tilde{s} = s\alpha_2\beta_1 = (q_+ + q_-)^2 + (\mathbf{q_1} + \mathbf{q_2})^2,$$

$$T_{\alpha\beta} = \gamma_\beta \frac{\hat{q}_1 - \hat{q}_+ + m}{(q_1 - q_+)^2 - m^2}\gamma_\alpha + \gamma_\alpha \frac{\hat{q}_2 - \hat{q}_+ + m}{(q_2 - q_+)^2 - m^2}\gamma_\beta.$$

(for details see [12]).

Generalization for the case of arbitrary number of exchanged photons between colliding nuclei is straightforward. The same approach as was used above leads to the following form of the generalized amplitude

$$M(2 \to 4) = \frac{1}{4}\int \frac{d^2\rho}{\pi}e^{-iq_1\rho - i\alpha Z_1 Z_2\psi(\rho)}\Phi_B(\rho, q_2), \qquad (30)$$

with

$$\Phi_B(\rho, q_2) = \int \frac{d^2\mathbf{k}}{\pi} e^{i\mathbf{k}\rho} M_p(k, q_2 - k).$$

Comparing the expression (30) with the amplitude for the process $3 \to 3$ (23) one can see that the crossing symmetry property between the considered processes takes place in the case when one neglects the multiple exchanges of particle C with nuclei. Moreover, this statement is correct even when one takes into account the screening effects between nuclei A_1 and A_2 in both processes, which manifest itself by insertion of light–by–light scattering blocks into Feynman amplitudes. As was shown in [13] accounting of this effect can be provided by the universal factor

$$\exp\left\{-\frac{\alpha^2 Z_1 Z_2}{2} LA(\rho)\right\}, \quad L = \ln(\gamma_1 \gamma_2), \tag{31}$$

with the complex quantity $A(\rho)$ connected with the Fourier transformation of light–by–light scattering amplitude.

Nevertheless, crossing symmetry is broken in all orders of perturbation theory if one tries to compare the full amplitude for the process $3 \to 3$ (expression (23) with the replacement (26)) and the relevant amplitude for the process $2 \to 4$ accounting the multiple interaction of produced particles [12].

Thus the crossing symmetry property takes place only for the colliding nuclei with charge numbers fulfilled the approximation $Z_{1,2}\alpha \ll 1$.

Acknowledgment

We are grateful to V. Serbo and N. Nikolaev for the useful discussions and comments. The work is supported by grants INTAS 97-30494, SR-2000.

References

1. L. D. Landau and E. M. Lifshits, Phys. Z. Sowjet. **6**, 244 (1934).
2. G. Racah, Nuovo Cimento **14**, 93 (1937).
3. A. J. Baltz, L. McLerran, Phys. Rev. C **58**, 1679 (1998).
4. A.J. Baltz, F. Gelis, L. McLerran, A. Peshier, Nucl. Phys. A **695**, 395 (2001).
5. R. N. Lee, A. I. Milstein, V. G. Serbo, hep-ph/0108014.
6. R. N. Lee, A. I. Milstein, Phys. Rev. A **61**, 032103 (2000).
7. U. Eichmann, J. Reinhardt and W. Greiner, Phys. Rev. A **59**, 1223 (1999).
8. U. Eichmann, J. Reinhardt and W. Greiner, Phys. Rev. A **61**, 062710 (2000).
9. A. J. Baltz et al., Nucl. Phys. A **695**, 395 (2001).
10. U. Eichmann et al., Phys. Rev. A **61**, 062710 (2000).
11. D. Yu. Ivanov, A. Schiller and V. G. Serbo, Phys. Lett. B **454**, 155 (1999).
12. E. Bartoš, S. R. Gevorkyan, E. A. Kuraev and N. N. Nikolaev, hep-ph/0109281.

13. E. Bartoš, S. R. Gevorkyan, E. A. Kuraev and N. N. Nikolaev, Phys. Lett. B **538**, 45 (2002).
14. E. A. Kuraev, L. N. Lipatov, M. I. Strikman, Soviet ZhETP **66**, 838 (1974).
15. I Ginzburg, S. Panfil, V. Serbo, Nucl. Phys. B **284**, 685 (1987).

E. BARTOŠ ET AL.

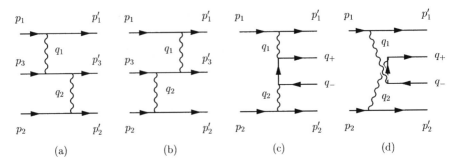

Figure 1. Feynman diagrams for Born amplitudes of the process $A_1 + A_2 + C \to A_1 + A_2 + C$ (a, b) and the process $A_1 + A_2 \to A_1 + A_2 + C + \bar{C}$ (c, d).

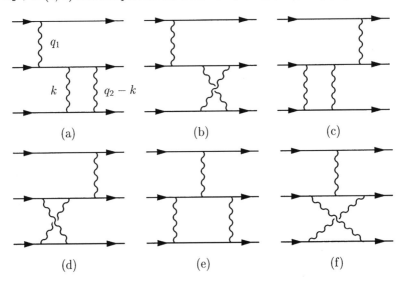

Figure 2. Feynman diagrams for the process $A_1 + A_2 + C \to A_1 + A_2 + C$ with three photon exchange.

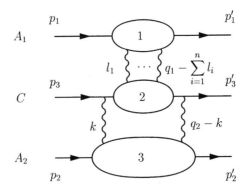

Figure 3. Feynman diagram for the process $A_1 + A_2 + C \to A_1 + A_2 + C$ with $n + 1$ exchanged photons ($n \geq 1$) between particles A_1 and C.

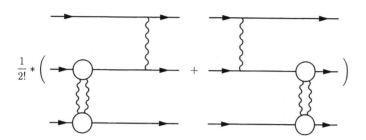

Figure 4.   Feynman diagram for the amplitude $M_{(2)}^{(1)}$.

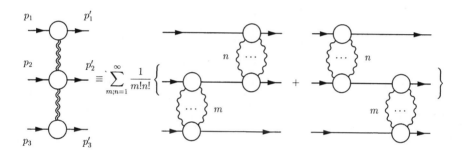

Figure 5.   Feynman diagram for the amplitude $M_{(\infty)}^{(\infty)}(q_1, q_2)$.

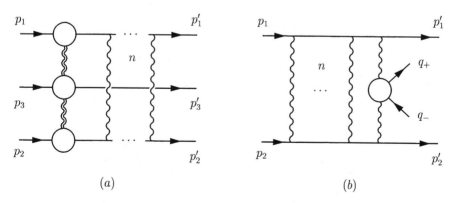

(a) (b)

Figure 6.   Feynman diagrams for the n photon exchange between nuclei A_1 and A_2 compared with the Born diagram for the processes $3 \to 3$ (a) and $2 \to 4$ (b) (blob in (b) correspond to diagrams in Fig. 1 (c, d)) .

List of Participants

A.Aghanyants
Yerevan Physics Institute
Alikhanian Brothers St.2, Yerevan 375036, Armenia
G.Aghuzumtsyan
Physikalisches Institut der Universität Bonn
Nussallee 12, 53115 Bonn, Germany
A.Airapetian
Yerevan Physics Institute
Alikhanian Brothers St.2, Yerevan 375036, Armenia
N.Akopov
Yerevan Physics Institute
Alikhanian Brothers St.2, Yerevan 375036, Armenia
Z.Akopov
Yerevan Physics Institute
Alikhanian Brothers St.2, Yerevan 375036, Armenia
M.Amarian
Yerevan Physics Institute
Alikhanian Brothers St.2, Yerevan 375036, Armenia
and
DESY-Zeuthen, 15738 Zeuthen, Germany
R.Avagian
Yerevan Physics Institute
Alikhanian Brothers St.2, Yerevan 375036, Armenia
H.Avakian
Jefferson Lab/ Hall B
12000 Jefferson Ave., Newport News, VA 23606, USA
A.Avetisian
Yerevan Physics Institute
Alikhanian Brothers St.2, Yerevan 375036, Armenia
E.Avetisian
INFN-Laboratori Nazionali di Frascati
Via E.Fermi 40, I-00044 Frascati, Italy
L.Baghdasaryan
Yerevan Physics Institute
Alikhanian Brothers St.2, Yerevan 375036, Armenia
N.Bianchi
INFN-Laboratori Nazionali di Frascati
Via E.Fermi 40, I-00044 Frascati, Italy
A.Borissov

Randall Laboratory of Physics, University of Michigan
Ann Arbor, MI 48109-1120, USA
P.E.Bosted
University of Massachusetts
Amherst, MA 01003, USA
A.Bruell
Massachusetts Institute for Technology
77 Massachusetts Avenue, Cambridge, MA 02139, USA
M.Danton
Schonland Research Institute, University of Witwatersrand
Johannesburg, Priv. Bag. 3, 2090, South Africa
P.Decowski
Department of Physics, Smith College
Northampton, MA 01063-1000, USA
K.Egian
Yerevan Physics Institute
Alikhanian Brothers St.2, Yerevan 375036, Armenia
G.Elbakyan
Yerevan Physics Institute
Alikhanian Brothers St.2, Yerevan 375036, Armenia
F.Ellinghaus
NIKHEF
PO Box 41882, NL-1009 DB, Amsterdam, Netherland
A.Freund
Institute of Theoretical Physics, University of Regensburg
Universitätsstr.31, 93040, Regensburg, Germany
S.Frullani
INFN, Gruppo Sanita and Instituto Superiore di Sanita, Physics Lab.
Viale Regina Elena 299, I-00161, Rom, Italy
S.Gevorkyan
Joint Institue for Nucelar Research
141980 Dubna, Moscow region, Russia
and
Yerevan Physics Institute, 375036, Yerevan, Armenia
V.Gharibyan
Yerevan Physics Institute
Alikhanian Brothers St.2, Yerevan 375036, Armenia
and
DESY, Notkestrasse 85, 22607 Hamburg, Germany
V.Ghazikhanian
UCLA Department of Physics and Astronomy
Box 951547, Los Angeles, CA 90095-1547, USA

O.Grebenyuk
Petersburg Nuclear Physics Institute
Gatchina, Leningrad District, 188350, Russia
M.Hovhanesyan
Yerevan Physics Institute
Alikhanian Brothers St.2, Yerevan 375036, Armenia
N.Ivanov
Yerevan Physics Institute
Alikhanian Brothers St.2, Yerevan 375036, Armenia
H.E.Jackson
Physics Division, Argonne National Laboratory
Bldg. 203, 9700, South Cass Ave., Argonne, IL 60439, USA
J.Kelly
Department of Physics, University of Maryland
College Park, MD 20742, USA
I.Keropyan
Yerevan Physics Institute
Alikhanian Brothers St.2, Yerevan 375036, Armenia
A.Kotzinian
Yerevan Physics Institute
Alikhanian Brothers St.2, Yerevan 375036, Armenia
and
JINR, 141980 Dubna, Moscow region, Russia
E.Kuraev
Joint Institue for Nucelar Research
141980 Dubna, Moscow region, Russia
A.Kuyumchyan
Institute of Microelectronics Technology
142432 Chernogolovka, Russia
G.K.Mallot
CERN
1211 Geneva 23, Switzerland
H.Marukyan
Yerevan Physics Institute
Alikhanian Brothers St.2, Yerevan 375036, Armenia
A.Miller
TRIUMPF
4004 Wesbrook Mall, Vancouver, BC V6T 2A3, Canada
W.D.Nowak
DESY Zeuthen
Plattanenallee 6, D-15738, Germany

A.G.Oganesian
Institute of Theoretical and Experimental Physics
B.Cheremushkinskaya 25, 117218 Moscow, Russia
D.Reggiani
INFN-Laboratori Nazionali di Frascati
Via E.Fermi 40, I-00044 Frascati, Italy
and
Universita di Ferrara, Ferrara, Italy
A.Rostomyan
DESY
Notkestrasse 85, 22607 Hamburg, Germany
D. Ryckbosch
University of Gent, Department of Subatomic Physics
9000, Gent, Belgium
C.Schill
INFN-Laboratori Nazionali di Frascati
Via E.Fermi 40, I-00044 Frascati, Italy
E.Steffens
Physikalisches Institut, University of Erlangen-Nürnberg
Erwin-Rommel Str.1, D-91058, Erlangen, Germany
S.Taroyan
Yerevan Physics Institute
Alikhanian Brothers St.2, Yerevan 375036, Armenia
H.Zohrabyan
Yerevan Physics Institute
Alikhanian Brothers St.2, Yerevan 375036, Armenia

List of Contributors

G.Aghuzumtsyan
Physikalisches Institut der Universität Bonn
Nussallee 12, 53115 Bonn, Germany
A.Airapetian
Yerevan Physics Institute
Alikhanian Brothers St.2, Yerevan 375036, Armenia
N.Akopov
Yerevan Physics Institute
Alikhanian Brothers St.2, Yerevan 375036, Armenia
M.Amarian
Yerevan Physics Institute
Alikhanian Brothers St.2, Yerevan 375036, Armenia
and
DESY-Zeuthen, 15738 Zeuthen, Germany
H.Avakian
Jefferson Lab/ Hall B
12000 Jefferson Ave., Newport News, VA 23606, USA
N.Bianchi
INFN-Laboratori Nazionali di Frascati
Via E.Fermi 40, I-00044 Frascati, Italy
A.Borissov
Randall Laboratory of Physics, University of Michigan
Ann Arbor, MI 48109-1120, USA
P.E.Bosted
University of Massachusetts
Amherst, MA 01003, USA
A.Bruell
Massachusetts Institute for Technology
77 Massachusetts Avenue, Cambridge, MA 02139, USA
A.Freund
Institute of Theoretical Physics, University of Regensburg
Universitätsstr.31, 93040, Regensburg, Germany
S.Frullani
INFN, Gruppo Sanita and Instituto Superiore di Sanita, Physics Lab.
Viale Regina Elena 299, I-00161, Rom, Italy
S.Gevorkyan
Joint Institue for Nucelar Research
141980 Dubna, Moscow region, Russia
and

Yerevan Physics Institute, 375036, Yerevan, Armenia
V.Gharibyan
Yerevan Physics Institute
Alikhanian Brothers St.2, Yerevan 375036, Armenia
and
DESY, Notkestrasse 85, 22607 Hamburg, Germany
O.Grebenyuk
Petersburg Nuclear Physics Institute
Gatchina, Leningrad District, 188350, Russia
N.Ivanov
Yerevan Physics Institute
Alikhanian Brothers St.2, Yerevan 375036, Armenia
H.E.Jackson
Physics Division, Argonne National Laboratory
Bldg. 203, 9700, South Cass Ave., Argonne, IL 60439, USA
J.Kelly
Department of Physics, University of Maryland
College Park, MD 20742, USA
A.Kotzinian
Yerevan Physics Institute
Alikhanian Brothers St.2, Yerevan 375036, Armenia
and
JINR, 141980 Dubna, Moscow region, Russia
E.Kuraev
Joint Institue for Nucelar Research
141980 Dubna, Moscow region, Russia
G.K.Mallot
CERN
1211 Geneva 23, Switzerland
W.D.Nowak
DESY Zeuthen
Plattanenallee 6, D-15738, Germany
A.G.Oganesian
Institute of Theoretical and Experimental Physics
B.Cheremushkinskaya 25, 117218 Moscow, Russia
D.Reggiani
INFN-Laboratori Nazionali di Frascati
Via E.Fermi 40, I-00044 Frascati, Italy
D. Ryckbosch
University of Gent, Department of Subatomic Physics
9000, Gent, Belgium